OLIVES

OLIVES

*Cooking with the olive
and its oil*

MARLENA SPIELER

CHARTWELL
BOOKS, INC

A QUINTET BOOK

Published by Chartwell Books
A Division of Book Sales, Inc.
114, Northfield Avenue
Edison, New Jersey 08837

This edition was produced for sale in the U.S.A.,
its territories and dependencies only.

ISBN 0-7858-0895-7

This book was designed and produced by
Quintet Publishing Limited
6 Blundell Street
London N7 9BH

Creative Director: Richard Dewing
Art Director: Clare Reynolds
Designer: James Lawrence
Project Editor: Clare Hubbard
Editor: Jane Hurd-Cosgrave
Photographer: Janine Hosegood
Food Stylist: Emma Patmore

Typeset in Great Britain by
Central Southern Typesetters, Eastbourne
Manufactured in Singapore by
Pica Colour Separation Overseas Pte. Ltd.
Printed in Singapore by
Star Standard Industries (Pte.) Ltd.

Picture Credits
pgs 6/7 International Olive Oil Council; pg 8 California Olive Industry;
pg 9 Life File; pg 10 (top) Alfredo Benavente Navarro, (bottom) Instituto
Sperimentale; pg 12 (top) Frantoio Gaziello; pg 14 (top) Olive Oil
Information Bureau; pg 16 Life File; pg 20 Life File.

Warning

Because of the risk of salmonella
poisoning, raw or lightly cooked
eggs should be avoided by the
elderly, the young, babies, pregnant
women, and those with an impaired
immune system.

Acknowledgments

One of the great joys of working on this book was the experience of being awash with olive oil. The other was meeting so many other people who felt the same—cooks, importers, fellow food writers—all as passionate about olives and their wonderful oil as I am.

I would like to thank all of the following for keeping me so supplied with olive oil: Panayis Manuelides (Odysea); Alice Seferiades (Odysea); the Monks of New Norcia, Australia; Rod Jeans (Messara); Elaine Ashton (Grania and Sarnia); Gianni and Pamela Parmigiani (whose appreciation and high quality of their olive oil is equaled only by their luscious Parmesan cheese); Lisboa Patisserie (Portuguese oil); Priscilla Carluccio (Carluccio's); David Roberts and Paolo Ardisson (John Burgess Exports); Charles Carey (Donatantonio Plc); Adrian Francis (Marks and Spencer); Carbonelli; Fresh Olive Co. of Provence; The Olive Press (California oil); Jon Eaton (The Beverly Mustard Co.); Mr. Silva and Mr. Gomes (Portuguese olive oil).

I would also like to give a special thanks to: Maria Jose Sevilla, and Foods and Wines of Spain, for olive-oil pressings and tapas throughout Spain; Mike Callaghan of Friends of the Olive Groves, Athens, Greece, for taking our group of food writers to Greece last summer, and bringing us to Mount Hymettus for the blessing of the olives; to Dr. Stefano Raimundi and the Italian Trade Commission for their help in all matters Italian; The Olive Oil Council (and all at Graylings PR who work with them, especially Caroline Black); Sally McCormick of the California Olive Oil Council; Edite Philips for her Portuguese olive-oil enthusiasm and help; to Anne Dolamore, Rosemary Barron, and Judy Ridgway for sharing their knowledge of the world of olive oil in their excellent books; Elaine Hallgarten for her samples of Israeli olive oil and to Richard Frenkel, Derek Monday, and everyone at Frenkel Oils, Ltd.

Nuñez de Prada invited me to an olive pressers' breakfast in Baena, and I shall never forget it (nor their salmoreja soup!). Schwartz Spices sent me a box of marvelous spices and mixtures, which found their fragrant way into many of the following recipes. Lynne Meikle at Tate and Lyle provided a wonderful sugar selection, which showed me just how good using olive oil to bake cakes and pastries with olive oil could be. I have also enjoyed using Le Creuset pots and saucepans increasingly every time I have discovered a new size and shape—their grill pan is the current star of my stove, perfect for preparing olive oil–scented, broiled dishes.

To Leah Spieler, always my little girl—even if she is grown up and almost a doctor. To John Harford for his enthusiasm regarding vegetables and everything, really. To Gretchen Spieler, just because.

To friends who ate: Rabbi Jason Gaber; Dr. Esther Novak and Rev. John Chendo; Sue Kreitzman; Sri and Roger Owen; Paula Eve Aspin and Richard Hudd; Amanda Hamilton and Tim Hemmeter; Sandy Waks; Kamala Friedman; Fred and Mary Barclay (who brought me olives from their home in Cyprus); Nigel Patrick and Graham Ketteringham; Jerome Freeman and Sheila Hannon; Paul Richardson; and my colleagues at the San Francisco Chronicle: Michael Bauer, Fran Irwin, M.A. Mariner, and Maria Cianci.

To my family: my parents, Izzy and Caroline Smith; Aunt and Uncle Sy; Estelle Opper; all my little cousins; and especially my grandmother, Sophia Dubowsky.

Thanks also to my very patient agent, Borra Garson; my olive-o-phile husband (who is also my publicist and manager); Alan McLaughlan; and that always-fine cat, Freud, without whom my kitchen would be a less-interesting place.

Preface

Resting in the shade on a very hot day, covered with a gaily colored cloth, stood a little table. On it was a plate of juicy, ripe tomatoes dripping with strong, fruity, olive oil and scattered with shiny black olives. There was also a rustic loaf of crisp-crusted bread, the sweet scent of summer in the air, and the lazy hum of bees.

I was about 17 years old when I sat in the Mediterranean afternoon sun, eating this lunch. The olive oil was rich and fragrant. It glistened on the tomatoes, on the bread, and on my fingers, which I could not stop licking. And the olives! Meaty, juicy, saline, slightly bitter, and utterly delicious. I had grown up in California eating mild (some call them insipid), "ripe" olives—but I had no idea that olives could have such incredible flavor! I

above *A variety of succulent olives.*

was smitten by this most ancient of fruit, and I have little doubt that the olive was instrumental in turning me away from my art education to point me in the direction of the kitchen.

When I returned to my California home, and began the journey to adulthood that starts by having your own apartment, I discovered something on the supermarket shelves: in a gold-colored can, with a very simple label, was some California-made olive oil. I bought it. It tasted green and grassy, and at once conjured up the olive-scented foods I had grown to love in the Mediterranean. I could not believe this wonderful fact: olive oil, grown and pressed locally, was on my supermarket shelves, and was also affordable! It had been there all along, only I hadn't noticed it.

My friends and I cooked and cooked, gave dinner parties, went to cooking classes, and traveled, always for the pleasure of eating. Olive oil was a large part of this cooking and eating extravaganza—it made everything taste so wonderful. Supermarkets, specialty shops, and delis seemed to be following us around, as they now began to devote shelves upon shelves to olive oil, where, before, there had only been one or two bottles. The next big discovery was the health benefits of the Mediterranean diet. Until then, most people still believed the major oil refiners' advertising, perceiving olive oil as an expensive indulgence, and a very unhealthy food. They claimed that margarine was more healthy. I instinctively knew differently—the people I had seen in the Mediterranean were so healthy, so active, and so very old, it couldn't be true. Suddenly, the Mediterranean diet was splashed throughout the press, and everyone was using an olive-oil diet for health purposes (and drinking red wine, too!).

Since then, my delight in the olive has continued to grow. It is a subject that fascinates me as much as it provides nourishment, since there is so much ritual and history packed into this healthy and pleasure-giving fruit. So much has my life revolved around the olive, that when my husband and I married, in a vineyard in California's Sonoma county, we planted an olive tree on the spot we exchanged our vows. And we are now awaiting our first crop!

Marlena

Introduction

Olives reveal more of themselves and offer enjoyment the more you taste and learn about them. Like wine, the preparation of olive oil is an ancient ritual of crushing the fruits and preparing them that is rooted in both soil and traditional custom.

Most Mediterranean lands eat olives and their oil with bread for breakfast, with bread and wine as a before-dinner snack, or (in Christian countries) with bread as a fasting meal for Lent. I am reminded of the monastery of Khrysorroyiatissa, north of Paphos on the island of Cyprus, where every visitor is given a traditional offering of bread, wine, and olives, each of which has been produced by the monks on the island for over 800 years.

Olives and their oil have also become an intrinsic part of the way we in the U.S. and Northern Europe eat, as these pair brilliantly with the savory flavors of garlic, chiles, vegetables, pastas, fish, and lighter meats. Olive oil delivers a great, satisfying taste that takes the place of other, richer foods, such as creamy, cheesy, or unduly sweet dishes. Olive oil is considered beneficial to health as it supposedly lowers the harmful low-density lipoprotein (LDL) cholesterol, while preserving the heart-protecting, high-density lipoprotein (HDL) cholesterol. Also, olive oil makes eating vegetables taste so wonderful that we eat more of them.

Olive oils, like fine wines, endlessly vary in color, flavor, and aroma. The reason for this wide variety of qualities in what is basically the same fruit from the same tree is a unique feature of the olive: regardless of which type was originally planted, over the many years of the tree's growth (and it can grow up to 600 years), it absorbs the flavors of its environment—the earth, the air, the winds and rains, indeed, the very soul of the people who cultivate it—which combine to create the final character of the individual fruit.

above **Young olives growing in a Californian grove.**

From the Ancient World to the New

The practice of cultivating the olive tree, *Olea europaea*, is so ancient that no one really knows when it first began. The olive tree has accompanied humanity's journey through history since prehistoric times, and is so intertwined with the stories of our past that it is impossible to ascertain who pressed the first oil or cured the first fruit.

The olive is believed to have been first cultivated in about 6,000 B.C. in what is now Syria and southern Iran, or possibly Greece. However, it was well known to the Egyptians, and seems to have been distributed throughout the central and western Mediterranean countries by Phoenician traders. It is often referred to in both the Old Testament and the Koran; other ancient people, too, attributed much of life's mystery to this potent, pungent fruit. In the days of the pharaohs, Ramses III planted olive trees around the town of Heliopolis, to provide oils to keep the lamps in the temples of Ra burning; also, three-thousand-year-old mummies have been found preserved with olive oil and surrounded by olive branches. Cured olives were often left in the pharoahs' tombs to provide food for the afterlife.

The Phoenician Greeks brought the first olive trees to Provence in about 6,000 B.C., and the Romans planted the first olive groves in Tunisia, where they remain a hugely important crop today (one in every eight Tunisians works in the olive-oil industry). The ancient Greeks believed that wisdom came from the olive; Plato is said to have walked his students through olive groves for inspiration, and local lore maintains that his tree still lives. In fact, near Athens, you will find a very old tree marked with a plaque designating it as "Plato's tree."

Olive oil has also retained a place in religious rituals of respect. If you visit Archbishop Makarios' tomb in the mountains of Cyprus, you can still see jars of the rich local green oil, left by the faithful as a mark of devotion.

Peace, too, was an offering of the tree: Greco-Roman mythology relays the story of the goddess Minerva striking the ground with her sword, which caused an olive tree to sprout, moving her to trade her implements of war for the olive branch of peace. The Old Testament (Genesis) describes the dove returning to Noah's Ark from Mount Ararat with an olive branch in its beak, showing the signs of hope as the flood waters subsided. Today, the olive branch is still used as a symbol of peace, and is in fact the symbol of the United Nations.

Excavations in Crete unearthed evidence that in 3,500 B.C., olives played an important part in the ordinary diet. Greek myth says that the goddess Athena created the first olive tree during a contest of creativity to come up with the most useful invention. Although Poseidon created the horse, the gods favored the generous, lyrical tree, and Athena was rewarded by Greece's capital being named after her.

By Roman times, olives represented a major part of both diet and commerce. The Romans ate olives as an antipasto, just like today, and believed olives to be an aphrodisiac. When the ruins of Pompeii were unearthed, crocks of preserved olives were found there. The Romans spread the gnarled, silver-green–leafed tree throughout the rest of the Mediterranean. By the time Rome fell, the olive tree was flourishing.

The importance of olives and their oil remained a largely Mediterranean discovery until 1560, when olive cuttings were taken to Peru by the Spaniards. In the 1700's, Franciscan priests brought the olive to Mexico, and northward along the mission trail that colonized

below *A tiled sign advertising olive oil in Cantabria Province, Spain.*

introduction

9

the area that then belonged to New Spain, then Mexico, and finally the U.S., when it became the state of California, where the olive industry is now booming. A famous Eastern Seaboard resident, Thomas Jefferson, once tried to grow olives at his home in Monticello, but the climate was too cold for them to flourish.

In the late 1800's, the olive was successfully transported to and cultivated in Australia, and shortly thereafter, was also introduced to South Africa by a fellow named Costa, I am told.

Many famous artists, among them Renoir, Van Gogh, Picasso, and Marcel Pagnol, all saw inspiration in the olive, immortalizing the gnarled tree and its fruit in their paintings.

Olive oil is one of the rare, happy stories in the progress of food technology and tastes. Whereas only a few years ago, olive oil was being neglected in favor of other cooking oils, most olive-producing areas of the world are currently celebrating their increasing olive growing and harvesting business, as more and more consumers yearn for the rediscovered qualities that olive oil uniquely provides.

Growing, Harvesting, and Pressing Olives

The olive tree prospers in areas where the summers are long, hot, and dry. It is an evergreen tree, and in addition to the hot summer, needs a winter cold enough to frost. The Mediterranean grows most of the world's olives, but many other countries or regions, including California, Argentina, South Africa, and Australia—even China—grow olives. And some of the olive oil from these lands is remarkable.

An olive tree needs five to six years to produce a respectable yield of fruit; the tree reaches its full fertility at about 20–25 years of age, when they can yield up to 45–90 pounds. Many olive trees only produce olives every other year, though some bear annually.

There are many, many varieties of olive tree, each with its own distinctive taste, scent, and oil level—some olives are rich and flavorful, but yield little oil. These are best for eating. Other olives are rich with oil, and these are the ones used for pressing. While the olive itself is determined by its botanical characteristics, it is formed, molded, and changed by the climate and soil of the lands where it is grown. The olive tree adapts readily, hence the wide array of flavors from the final oil.

The tree blossoms in May and June, and then produces tiny berries that grow to the fleshy, hard-pitted fruit we are familiar with. Harvesting can take place anywhere from late October to February, depending on the weather of the region of cultivation. The olive ripens and changes colors, from green, to purplish, then reddish, then black; at this point, most olives have their maximum amount of oil. It takes

above **Picking olives in the Abruzzo region of Italy.**

below **A rugged, sun-baked grove in Sierra de Segura, Spain.**

introduction

about 11 pounds of olives to produce 1¾ pints of oil, and each tree can produce 5¼–6 pints of oil per year.

The timing of the harvest determines the oil's flavor. Each variety has its peak of maturity, when that particular olive is best for its oil. In Tuscany, olive picking generally starts in October, often the day after the end of the grape harvest, when the olives are green, firm, and lightly underripe. In Spain, November is harvest time. In Greece and Cyprus, some are picked early, when green, but others are left to ripen to an oily black, often still being picked as late as February.

Underripe fruit gives a different flavor—too green, and you risk too much bitterness; but the right level of underripeness of some olives yields the best, bitter edge. Leaving the olives on the tree until they are ripe or overripe, harvesting as late as the middle of winter—even from olives that lie on the ground having fallen off on their own accord—gives some olives a stronger taste, yet others a too-high acidity level. In Portugal, olives are picked when very ripe. This may be an acquired taste, yet it seems a compelling one.

While fresh olives seem robust and hearty, they are in fact easily bruised, and when injured, develop bad flavors. Hand-picking gives a superior result, with lower acid and cleaner flavor, but it is very expensive. Greece is famous for the frequent sight of old ladies dressed in black, beating the trees so that the olives fall into cloths set on the ground. Less picturesque are the workers combing them onto nets hung from the trees.

In some places, olives are harvested in a more passive way: nets are stretched on the ground to catch the olives as they fall, and the nets are emptied every so often. This is convenient, but the cloths must be emptied often, as the olives start to oxidize as soon as they leave the tree, and in warm climates, begin to deteriorate immediately. Generally, three days is the maximum—any longer, and the olives ferment or oxidize, giving the oil unpleasant flavors and defects.

There are two methods of extracting oil from olives. The first is the *traditional* way: the olives are washed once, then pulped in a revolving mill, with the paste spread on mats. Then the mats, with their olive-purée spread, are piled up and pressed. The liquid that collects contains both water and olive oil, with some impurities and debris. Placed into a centifuge or evaporated, the oil of the olive appears—cold-pressed, virgin oil, tasting of thick, fresh olives. This is called "cold pressing."

Modern extraction washes the olives up to four times, then crushes the fruit into a paste using stainless-steel blades. This thick, sludgy paste of oil, water, and solids is then placed into centrifuges at a temperature of about 68°F, which separates the slightly warm oil from the water and solids. A temperature higher than 86°F will damage the oil; it must remain lower to be considered cold-pressed. (By the way, no other vegetable oil is edible by just being pressed—the other seeds, grains, etc., all need to be filtered and processed in some way.)

Quotes and Proverbs about Olives

(Spanish) *"Como la oliva verde es el querer que te tengo sumiso y blando por fuera, entero y fuerte por dentro."* "My love for you is like the green olive, yielding and soft on the outside, sound and strong on the inside."

(Provence) Today's rain brings tomorrow's olive oil.

(Italy) *"Olio nuovo e vino vecchio."*—New oil and old wine.

(Greek) Welcome, little olive, and your sweet, sweet oil (from a Corfiot folk song).

In France and Greece, the final pressing of olive oil, which is too inferior to cook with, goes to make a lovely, green, olive-scented soap.

In the Mediterranean, olive trees dot the landscape, and many families have several on their land or in their gardens, growing just enough for their own consumption—jars of brined olives, baskets of salted olives, and bottles of the fragrant oil. When I lived in Israel, a band of women from the nearby Arab village came to our house each winter when the olives were ready for picking, went away with great basketfuls, and returned a week later with crocks of preserved olives and bottles of oil.

Olive Oil-Based Cuisines

Throughout the Mediterranean, the rich, heady scent of olive oil permeates the entire dish of every cuisine, while markets offer a dazzling array of preserved olives, sold as great crocks filled with a painter's palette of colors, shapes, and flavors. From the salads and *daubes* of Provence to the zesty pastas of Italy, from the intricately spicy North African dishes to the exotic Middle-Eastern stews and roasts, grills, sandwiches, and street snacks, olive oil forms the most important flavor.

Indeed, in many countries, most breakfasts consist of bread and olive oil in various forms. In Spain, these are served with tomatoes and garlic; in Israel and the surrounding Arab countries, with pita bread and a spicy mixture called *dukkah*; in Italy, magazine advertisements show a mother drizzling olive oil onto bread for her children, reading "children need love … and olive oil."

But this passion for olive oil is not wholly Mediterranean, as the New World has also embraced the olive—today, modern American cuisine, particularly in California, is emphatically olive-oil–based, in much the same way as Mediterranean food. In northern California, you'll often find a saucer of olive oil along with your bread instead of butter, as the olive trees flourish throughout much of the Mediterranean-type climate of the state.

In South American countries, which also received plantings brought by the Spanish, there are examples of olive-oil-flavored dishes. Peru has a number of olive-oil–flavored dishes, especially the luscious *causa a la lemono*, a lemon-chile mashed-potato salad largely flavored with olive oil. Argentina also grows olives and presses their oil, though they seldom export it now. I have high hopes for tasting some of their oils in the

above ***The modern centrifugal plant at the Frantoio Gaziello company, Liguria, Italy.***

above ***The quintessential Mediterranean breakfast—fresh bread drizzled with olive oil.***

List of Delicious Things to do with Olive Oil

- Use olive oil—a mixture of pure and extra-virgin oils—for making French fries.
- Mix extra-virgin olive oil into garlic butter.
- Fry eggs in a small amount of extra-virgin olive oil (with cumin, paprika, etc.).
- On bread—chilled until it is firm as butter. In Provence, in deliciously rough places like Marseilles, a very fine, flavorful, extra-virgin olive oil is traditionally chilled until firm, then spread onto bread.
- Stir extra-virgin olive oil into soup: especially in Tuscan-style bean-and-pasta soups, such as the robust potage of garbanzos, pasta, and vegetables known as *frantoio*, or oil-pressers' soup.
- Drizzle extra-virgin olive oil onto bread, then sprinkle with garlic, diced tomatoes, and a crumbling of goat's or feta cheese, sprinkle with oregano, and eat for a Mediterranean-inspired breakfast.
- Cook an omelet or scrambled eggs in extra-virgin olive oil—add diced feta or goat's cheese, a handful of spinach leaves, a few sliced, wild mushrooms, or sautéed zucchini slices.
- Deep-fry squid, potatoes, chicken, croquettes, or anything you like crisply fried, using pure olive oil with a little extra-virgin olive oil.
- Use olive oil to brown French toast or pancakes.

future. Chile and South Africa, too, are developing olive-oil producers. Another promising producer is Australia. The oil produced by monks in New Norcia, is one of my current favorites; a few drops of this oil gives a startlingly delicious hint of olive to any dish. Wonderful!

Enjoying and Storing Olive Oil

Store olive oil in a cool cupboard away from direct light; green glass helps to protect the oil from oxidization. Unopened and properly stored, olive oil should last 18–24 months, though its flavor fades from the moment it is pressed. Once the bottle is opened, the fragrance and flavor begins to pale, and rancidity occurs from six months to a year.

Olive oil does not need refrigeration. If refrigerated, the oil becomes thick and almost solid, opaque, and cloudy. This does not harm the oil though, which returns to a liquid state when it warms up. However, it is hard to pour when it is so thick. In Provence, olive oil is sometimes chilled or even frozen, and enjoyed as a spread on crusty bread.

Olive oil is excellent for frying, as it remains stable at high temperatures—415°F which is higher than the 315°F of corn oil or the 325°F of sunflower oil. Pure olive oil is good for frying—it is less expensive than extra-virgin oil. A good-quality, pure olive oil should have an olive scent when heated; if it doesn't smell enough like olives, add a generous amount of extra-virgin oil to the frying pan along with the pure oil.

Though flavoring oils with garlic and herbs remains popular, and is recommended in many cookery books and articles, it is in fact a dangerous thing to do, as the oil seals in the ingredients, making it a perfect culture-growing medium for botulism and other bacteria. If you like flavored oils it is best to buy them from the store. Commercially prepared oils are safer, since they have been sterilized and treated to discourage the growth of bacteria.

Tasting and Choosing Olive Oils

The array of flavor characteristics is wide and varied. Like art, music, literature, and other civilized pleasures, olive oil yields its rich variety the more familiar you are with it. There are over 60 different types of olives grown in the world, some with a higher oil content, others with more flesh. The type of olive you have will dictate what you do with it: Kalamata olives, for instance, are delicious brined, but if you try to press them, you will have a disappointingly small yield. The Kalamata olive oil you buy in stores is oil from the Kalamata region, but it is Koroneiki, a different type of olives from the Kalamata enjoyed at table. Greek olive oil has a leafy quality, which I thought odd until I discovered that the leaves are pressed with the fruit, which contributes to the leafy aroma.

Tasting a few olive oils together informally can alert you to their variety, though if you taste more than five to six at a time, you will only be confused. For a successful, informal tasting session, purchase several

different types of olive oils, and set out a bowl of bread cubes and a plate of sliced apples along with saucers of the oils. This is not scientific, by any means, rather a social and interesting way to approach tasting the oils. You might like to add plates of thinly sliced vegetables, such as tomatoes, fennel, red peppers, and artichoke hearts, for a salad-type feast, but for strictly tasting the oils, omit the salad.

Professional olive oil tasting, however, is serious business—when the olives are picked and pressed, the oil that is exuded is not necessarily perfect. There are a wide variety of defects, any one of which can ruin the oil. If the olives are bruised, the oil's acidity may be too high. If the olives have molded, the oil will be musty. Therefore, all olive-oil producers have olive oil tasters, able to identify unpleasant flavors and smells from various batches. The vocabulary of tasting olive oil is similar to that of tasting wine (or beer, coffee, tea, or indeed any complex liquid). Nuances, fragrances, color, bouquet, flavors, and aftertastes are all parts essential to the whole quality of the oil. There are also numerous laboratory tests put to the olive oil to test its composition.

The consumer, though, is in a more difficult position. You want an oil that gives you pleasure, that you can use with abandon, rather than an expensive luxury to dispense with each miserly drop. Olive oil is a commodity, a life-enhancing food, meant to be generously drizzled over salads and splashed into soups, dripped onto bread, and even made into cakes. Therefore, you want an oil that is reasonably priced.

Choosing the best-quality and best-value oil is not such an easy task. While you can appreciate the occasional, exquisite treat of a very, very, fine oil, it is the everyday, good oil that gives its distinctive personality to your food. There should be no elitism involved in choosing an olive oil.

above **Choose your oil carefully, as there is a vast array of products available.**

below **Labels used on the extra-virgin olive oil produced by Fattoria Dell'Ugo, Italy.**

Fattoria dell' Ugo

Olio Extra Vergine di Oliva di Frantoio
Prodotto e Imbottigliato dalla
Fattoria dell' Ugo
di Amici Grossi
Tavarnelle Val di Pesa-Firenze
PRODUCT OF ITALY
Contenuto Netto 0,375 Litri

NON DISPERDERE IL VETRO NELL'AMBIENTE

Da consumarsi preferibilmente entro Giugno | 1994 | 1995 | 1996 | 1997 |

Don't be swayed by designer labels—olive oil is not a fashion statement, it is a food. While many of the fashionable oils are good, they are often priced as much as five times their worth. And they are not always superior—a rather depressing point when you have just forked out $15–$20.00 a bottle! Most of the olive oils—90 percent—sold in this country are sold in supermarkets. Until recently, most of this has been inferior, rather insipid olive oil. Happily, this is now changing.

To taste olive oils like the professionals, do it first thing in the morning—before breakfast, and definitely before coffee. Do not shower with soap, and do not dab yourself with perfume, so you will be able to smell the oil clearly. Give the bottle a good sniff, then pour it into a spoon. Roll the oil in your mouth, aerating it a bit, as you would wine. Then, spit it out. You should have flavors remaining on your palate, like the peppery aftertaste of Tuscan oil, or perhaps an almond-like essence or a wood scent. Try to taste the lightest oil first, and then work your way to the strongest, as with wine. Many say that you should not spit, as the main sensations of taste continue around the top and sides of the mouth and tongue, and down into the throat. You might want to perform a combination of swishing, spitting, and swallowing.

The tasters' vocabulary, officially drawn up by the International Olive Oil Council for professional tasters, is described here briefly. The following are styles of oils: "aggressive" or "pungent" refers to strong, initial flavors and/or aromas; "bitter" is self-explanatory, and can be either pleasant or not. "Delicate," "fresh," "fruity," "strong," "rustic," and "harmonious" are all self-explanatory; "sweet" means pleasant, and neither bitter, astringent, or pungent.

Aromas and flavors range from the following: "fruity," such as apple, banana, lychee, melon, pear, tomato, and olive; "verdant" includes eucalyptus, flowers, grass, leaves, hay, herbs, mint, violets; "vegetal" includes avocado; "earthy," "rustic," "nutty" refer to scents of almond, brazil nut, or walnut; and some oils are even found to be "chocolatey!" Taste for yourself: when the first signs of identifying specific aromas and flavors reach your palate, you just might find yourself squealing with delight as I did.

Terms and Types of Olive Oils

Extra-virgin is pressed from the fruit of the olive tree by crushing, without the use of chemicals or heat, and will have a maximum free acidity of 1%. First cold-pressing is still used as a description, since consumers are familiar with this term, but all extra-virgin olive oil these days is from the first cold-pressing. Whichever country your extra-virgin olive oil comes from, there are three different levels of quality and character.

Mainstream oils are the sort of oils you'll find in supermarkets, often own-brands, and they can be excellent value. Often the quality is superb, depending on who is supplying the oil.

Fine virgin is no more than 1.5% oleic acid.

Virgin is pressed from the olives without heat or chemicals, with a maximum free acidity of 2%.

Semi-fine is no more than 3% oleic acid.

Regional oils have the distinctive character of the area in which the olives were grown, with identifiable aromas and flavor nuances. Regional oils will often be made from the olives of numerous mills in the area.

Estate oils are produced from oils grown on one particular farm. The olives will usually be hand-picked and bottled on site. They are expensive and usually, exquisite.

Olive (also known as Pure Olive) is a blend of olive oil which has been refined, then mixed with a percentage of extra-virgin olive oil. Some pure olive oils can be quite good, it depends on the amount and quality of the extra-virgin oil added. You can add your own extra-virgin oil to make a balance you enjoy; the ratio is usually around 85% refined oil to 15% extra-virgin. It should have a maximum free acidity of 1.5%.

Olive pomace is a blend of refined pomace olive oil with virgin olive oil. The olive pomace oil is extracted by chemicals from olive residues and then refined. It is also mixed in the same proportions as pure olive oil and has the same maximum free acidity.

Light olive oil is olive oil with a very small amount of extra-virgin oil added. There are also blends of vegetable and olive oils, including ones that have absolutely no extra-virgin oil added but are promoted as "The Mediterranean Diet." Do not be fooled.

above *An aging olive tree with a split trunk, Alicante Province, Spain.*

The Best Uses for Olive Oil in Cooking

The answer to the question, "When is it best to use olive oil in cooking?" is—always! Few things that are delicious are not improved by olive oil, even many cakes and biscuits.

Many people, including those who should know better, suggest that for cooking one should use the pure-grade olive oil, and save the extra-virgin for salads or for flavorings. While pure olive oil, mixed with a bit of extra-virgin, is a good medium for deep-frying, and more economical as well, and homemade mayonnaise can be a little too oppressive with extra-virgin olive oil, emphatically, I would still recommend using extra-virgin olive oil for everything. Cook a selection of vegetables in extra-virgin olive oil, and you have a feast. It's that simple. I have heard people, even food professionals say: "The flavor of the oil dissipates in the cooking," but this is not entirely true—the flavor perfumes the dish, and while it loses sight of itself perhaps, it becomes one with the food it cooks with. Others have said that they don't want the strong flavor of extra-virgin oil when they are cooking, and though it is true, this is a matter of personal taste—I, for one, find its strong delicious scent and flavor exceptionally alluring.

For those who have grown up in the Mediterranean, using olive oil in foods is second nature. They are used to eating in the Mediterranean style, with tomatoes dripping in olive oil, olive oil on bread, or oil drizzled into soup instead of cream, or oil-cooked vegetables, eggs, meat, and fish. Life without it is unthinkable. They do not need to learn how and when to use it, they have grown up with it since childhood. Those of us who did not grow up in the Mediterranean need to be told how to choose and when to use olive oil, as we did not absorb it from our surrounding culture as a child.

A good way to stock your kitchen is to keep about three different olive oils in your kitchen at any time. One high-quality, cold-pressed varietal or single-estate oil, for drizzling over salads, roasted vegetables, pastas, or bread and tomatoes. A cheap, well-made, commercial, extra-virgin oil to cook vegetables, sauté chicken, or brush onto fish, or cook tomato-based sauces with. Then you might consider a pure oil or a

below **Stock your kitchen with a variety of oils.**

mixture of pure and commercial-quality oils for deep frying, or any time you want a milder olive flavor. Often, I omit the last oil, and keep two different commercial, extra-virgin oils instead, using them for all of my cooking needs. Potatoes fried in extra-virgin olive oil are one of life's most delicious foods. As you develop your taste you will become more familiar with particular areas that produce oils you favor. Traveling broadens your palate with olive oils, and so often when I drizzle a little Tuscan, Greek, or Spanish olive oil over a piece of bread at home, I am immediately transported by memories of vacations or travels by the scent and flavor.

Olive Oils and Health

The role of fats in our diets is a contradictory one. The results are not yet in, but research continues to suggest that a diet rich in olive oil is healthy and can even lower cholesterol levels. Olive oil is high in vitamin E and anti-oxidants, which help keep the body's cells from aging, and being a mono-unsaturated fat, it also encourages a greater level of the high-density lipoproteins (HDL) that protect your arteries, while discouraging the formation of low-density lipoproteins (LDL) which deposits more cholesterol in the arteries. The increase in HDL may also help to prevent cholesterol deposits. In areas where the people eat a "Mediterranean diet" high in olive oil with lots of vegetables and fish, and little animal fats, there is considerably less coronary disease as well as breast and bowel cancers.

Olive oil contains two of the essential fatty acids, oleic and linolenic acid, which the body cannot make itself. (Oleic acid is also found in mothers' milk, so it is no surprise that it aids normal bone growth.) Combined with its supply of vitamins E and A, it has the most balanced composition of fatty acids of any vegetable oil.

17

Olive Oils From Around the World

Labels are occasionally misleading. Though most bottled olive oils have labels stating "bottled in Italy," that does not mean that the olives were grown there. Similarly, if it says "pressed in—," the olives could have been grown in one country, pressed in another, and bottled in yet another. And olives can come from different sources, too. There is no problem with this; it does not necessarily indicate inferior quality. It can, however, be confusing. Blended olive oils, like blended wines, can be good oils, but like wines, they are strongly affected by their geographical character—for example, the strong, delicious flavors of Greece, Tuscany, or Tunisia, are each quite distinct.

FRANCE French olive oils tend to be light, sweet, fragrant, and pale-colored, pressed from the *Tanche* that grow in Nyons, and the *picholine* closer to the coast. French olive oil is grown and pressed on a small scale, and in general is of a very high quality. Several of my current favorites include L'Olivier (from Nice—a bright-yellow color with no hint of green, yet a delightfully assertive olive flavor, and Le Vieux Moulin (from the Nyons region): crisp and almost apple-like, with a sweet edge. Then there is Moulin Alziari, also from Nice—when you enter the little Alziari shop in Nice, you are surrounded by a world of olives, and barrels of good, sweet oil. It has an aroma of dried fruit that almost echoes the taste of asparagus.
Recommended Oils: The Fresh Olive Company of Provence, Huile d'Olive, Alziari, Emile Noel Organic.

SPAIN Spanish oils remind me of that old children's rhyme: "When she was good, she was very, very good, but when she was bad, she was horrid!" Actually, when Spanish olive oils are bad, they are simply dull (as are any, to be fair).

There are four established, demarcated regions of olive-oil production. Catalunyan oils tend to be appealing, fruity, almond-scented, and vibrant. Borjas Blancas and Siurana in Catalonia press mostly Arbequina olives, and Sierra de Segura and Baena in Andalucia press Picual olives. But even the oils from outside these areas can be very good indeed. It is the mass market labels that have led me astray.

In Baena, Nuñez de Prado's traditionally pressed oil is hand-bottled, lyrically delicious, and ripe with sensual flavors that conjure up tropical winds and exotic essences.
Recommended oils: Nuñez de Prado of Baena, Costa d'Oro, L'estornell organic, Carbonnel, Coumela.

ITALY For many, Italy is synonymous with olive oil, and indeed there is a huge variety of olives grown and pressed for oil throughout the many regions of that country. Tuscan oils, especially from Lucca, have a reputation for being the best in Italy, and—many say—the world. With its peppery aftertaste and herbal, grassy flavors, Tuscan is the definitive olive oil.

above **Alziari extra-virgin olive oil is made from the tiny Caillette olive, producing an excellent oil.**

below **Cultivation on the Nuñez de Prado estate is totally organic.**

introduction

Many of the other regions also have distinctive oils that are certainly more reasonably priced than the fashionable Tuscan oils. Ligurian oil, for example, is as light and vivacious as those from Provence, fruity Umbrian oils are sweeter and less bitter than their Tuscan counterparts, and Molise oil, which is fresh and grassy, with a slow, subtle, peppery aftertaste. My favorite from Apulia, Masseria—a rich, ripe, deep olive-flavored oil—is also a worthy variation.

Many Italian oils are in fact imported to Italy from Spain and Greece; this is not a criticism, as many of these oils are good. It is simply that mixing them obscures the geographical reference and character, and even if the oil is delicious, it is like a wine that has been blended rather than a varietal. Steer clear, however, of the mass-produced oils, as these can be disappointing.

Recommended oils: Novello, Santa Sabatina, Azienda Olearia del Chianti, Petro Coricelli, Clemente (la Zagare), Masseria.

GREEK Greeks consume more olive oil per person than anyone else on earth. Their extra-virgin oils are strongly flavored, rustic-tasting and bursting with Mediterranean energy. The Peloponnesus and Crete are the two biggest olive-oil producers, and their oils are mostly derived from the Koroneiki olive.

My personal favorite is Messara, though I recently experienced one bottle that thrilled me, while the other was insipidly flavored. Ditto for Karyatis. However, both were fantastic when fresh.

If I had my choice, I would happily taste my way through various olive-oil flavors, but inevitably drift back to good Greek oils—perhaps because the time I spent on Crete enabled me to appreciate the lusty, full flavor of the olives.

Recommended oils: Messara, Solon, Karyatis, Kolymvari, Iliada.

PORTUGAL Portuguese oils tend to be gold-colored, strongly flavored, a bit overripe, and though many are less than wonderful, I have often found them to be very useful in cooking. Sometimes a Portuguese oil might not hold up well in a proper tasting situation, yet when splashed onto a plate of tomatoes or a piece of fire-roasted fish, its deliciousness shines! Portuguese oils also tend to be quite reasonably priced.

Recommended oils: Quinta de la Rosa, "Gallo" Victor Guedes.

USA (CALIFORNIA) Having spent much of my life buying my olive oil directly from the pressers, bringing my bottles and having the fragrant oil decanted directly into them, I found it interesting to compare Californian varieties with traditional Mediterranean oils. California oils tend to be fresh and light; though they sometimes have a rougher feel, they range from okay to very, very good, and the industry is improving through experimentation with different types of olive.

Recommended oils: Bariani, Napa Valley Olive Oil, The Olive Press (Glen Ellen), Conzorzio.

above **The olives used to make Iliada oil are harvested from groves situated around the city of Kalamata.**

ISRAEL The best way to enjoy Israeli olive oil is to take an empty bottle to the nearest Arab village to have it filled with cold-pressed oil. In Haifa, you will find a museum devoted to all sorts of edible oils, with a strong accent on olive oil. Exhibits cover the ancient methods of olive pressing to the modern techniques. Land of Canaan makes a lovely little oil, quite golden in color, fruity, and almond-y; there are also others, such as the oil put out by the Golan Winery, which will hopefully begin to be more available to those of us in the rest of the world.
Recommended oil: Land of Canaan.

TURKEY Ranking fifth in the world's olive-oil–producing countries, Turkey not only produces a huge amount of olive oil, its cuisine is entirely olive-oil based (Turkish folklore tells of the *imam* [priest] in the eggplant dish, *imam bayildi*, who ate so much olive oil he swooned).
Recommended oil: Beverly Mustard Co., Kristal.

AUSTRALIA This was the surprise delight of this book! New Norcia olive oil, made by monks in Western Australia, is a sublime little oil. Seek it out.
Recommended oil: New Norcia. Unusual for its freshness and full flavor. Highly recommended, if you can get a bottle.

LEBANON Lebanese olive oil is beginning to make its way into our markets. I tasted a home-pressed, unlabeled sample; it was delicate and light, but not insipid, and reminded me a bit of Turkish olive oil at its best. I have high hopes for the oils from this area, as political tensions are easing and life is getting back to normal.

TUNISIA As the largest producer in North Africa, the olive is extremely important to the Tunisian economy, with more than 20% of the population working in the industry. About half the oil is exported to non-European Union countries.

MOROCCO AND LIBYA These countries also produce substantial amounts of olive oil, and enjoy a healthy cuisine based on it.

ARGENTINA AND CHILE Both of these South American countries have a substantial olive-growing and pressing industry, but at this point in time neither exports widely and I was therefore unable to taste any for this book.

NEW ZEALAND Produces small amount of high quality oil.

SOUTH AFRICA Production here is on a small scale, but the oil produced is of a high standard.

opposite *Different varieties of olives, served simply, are a perfect appetizer.*

below *A mouthwatering display of olives on a French market stall.*

introduction

Curing and Different Types of Olives

The bitter flavor of the olive is caused by a substance called oleuropein. To be edible, the oleuropein needs to be removed by curing. The curing process also softens and preserves the olives, releasing their flavor. There are several ways of curing olives. The one chosen usually depends on the tradition of the geographical area in which the olives are picked, and also the type of olive you are dealing with.

Spanish tradition says that the first edible olives were discovered near the sea, the bitter fruit having fallen in and been cured by the bath of the salt water in the rock pools, much like the modern brine that is used throughout the Mediterranean. The herbs that grow on seaside rock—rosemary, wild fennel, marjoram, thyme, and sage—added their herb flavors and scent, especially when gently heated by the sun, much like the marinades you often find olives bathed in. Essentially, this is the brine bath that most modern Mediterranean olives are cured in.

Another way to cure olives is to use a lye (in California) or a wood-ash (in Spain) bath. This gets rid of the oleuropein, but tends to dull the flavor, too. The resulting olive is the California "ripe" or the mild olives sold in cans from Spain. Sometimes, olives are cured by a combination of the two methods, a brine soak after a lye bath.

A third way is to layer them with salt, and when their bitter juices have leached out, to pack them in olive oil or store them dry and lightly salted. I have only seen this done with black olives.

Throughout the olive-growing regions of the world, curing and seasoning olives is an art, a tradition, an essential part of culture and dining. And the variety is nearly endless. Some olives are tiny nuggets of strongly delicious, flavored flesh such as Niçoise, others are fat, green Gordals or Queens, nearly as big as hens' eggs, and full of juicy flavor.

In addition to the variety of grown olives, they may also be cured in salt or brine, soaked, or dry and wrinkled. There is also a wide array of marinades: green olives awash in harissa (a fiery-hot sauce from Tunisia) or green olives studded with lemon and cracked coriander seeds, green olives tossed with sun-dried tomatoes, and black olives cloaked in olive oil and oregano, or with onions, pimientos, and carrots. Or they might be stuffed—black olives with almonds, green olives with anchovies, with pimiento, or with my favorite: whole cloves of garlic—not to be missed!

The names of olives can be based on either the cure, the marinade after the cure, or the type of olive or its place of origin, so choosing an olive by its name has no rationale; just remember an olive you particularly liked, and purchase it again next time.

There are so many varieties of olive that I could not recount them all. Instead, I have streamlined down to a useful few, as follows. The only definite advice I can give is this: do not buy olives that have been pitted (unless they are already stuffed); there is an indefinable flavor sensation that remains on the pit, as well as the pleasure of chewing on or sucking the pit after the flesh of the olive has been consumed.

The Blessing of the Olives at the Monastery of Mt. Hymettus

One hot June day, I joined a group of food writers at the Monastery of Mt. Hymettus, Greece, for the monks' blessing of the olive oil. Surrounded by the lushness of the forest, the monks spoke to us of the richness that olive oil gives us, admonishing us not to worship the oil itself, but the divine forces that created this wonderful fruit, plucked from the garden we call life.
We then tasted a range of oils. Each oil slowly yielded its differing and richly varied flavor, as we dipped into it with our chunks of breads, tasting each and moving around the table to taste the next sample. I was fascinated with the way the oils changed as I tasted them. Then lunch was served: a vegetarian feast of olive-oil cooked foods—Summer Squash Casserole (pg. 74), ripe tomatoes and feta cheese, black-eyed-pea salad with capers, and pasta with octopus—all tasting deeply flavorful and fresh, redolent of rich olive oil.

introduction

Amphissa: Firm outside, tender inside, plump, and reminiscent of both grapes and sea.

Arbiquena: Smallish, green-brown, often with leaves still attached, and fresh-tasting rather than aged and ripened.

Black Cerignola: Shiny and inky black, this has the texture of a ripe California or a canned Spanish olive, with a brinier, more olive-like taste.

Elitses: Tiny, Greek, table olives with a tangy astringency. They are brined, and range in color from olive-green–brown to black, often in the same batch of olives. Similar to Niçoise and Ligurian olives.

Greek Style: Big, fat, purplish-black, juicy olives, with lots of flesh and flavor. Sold brined.

Green and cracked: Slightly bitter, cracked, and brined, sometimes marinated with coriander seeds, lemon, cumin, or other flavorings. Naphlion is one of the most famous of this type.

above **Greek-style olives.**

Ionian: Plump and fleshy, these green olives are from Southern Greece, and are best in salads and used as appetizers.

Kalamata: Oval and fleshy, with a slightly pointed end, the Kalamata is perhaps the world's best-known olive, and justifiably so. Its flavor and texture are superb, and it is delicious in nearly any place where a strongly flavored, tangy, black olive is required.

Manzanilla: The typically Spanish olive, which is plump and juicy, green in color, available both pressed and cured; originally exported to the New World from Spain.

above **Kalamata olives.**

Moulin de Daudet: The black Moulin de Daudet olives are oil-cured in the Provençal village of Fonvielle. They have a rich, deep flavor with a hint of anise. The green olives have a rich, buttery texture, with hints of herbs, pine, and citrus.

Niçoise: Tiny, herby, gray-green–to–black olives with a great flavor that conjures up the taste of Nice in every bite. Their more formal name is Cailletier.

Nyons: Large, delicious, black olives, famous throughout France. Their pleasantly bitter edge makes an olive that is good, whether brine- or salt-cured.

Oil or Salt Cured: Wrinkled black olives with a fruity flavor and a dry-ish, chewy texture, these range from having a slightly truffle-like scent to a distinctly bitter one. Some are leathery and tough, which can be unpleasant when eaten as is, but this leatheriness is a boon when they are simmered into stews and sauces, as the flesh grows more tender and the bitterness adds a note of depth. Gaeta olives are one of the most-often encountered, cured type.

Picholine: The main green French olive, small, long, and yellow-to-green in color, often only lightly cured, which gives it a fresh, lively flavor.

Queen or Gordal: Huge, green olives, fleshy, and nearly the size of small hens' eggs.

Royal olives: Huge, fleshy olives, greenish-purple, with a robust flavor.

A Note About the Following Recipes

I intended to offer classic, olive-oil–scented or flavored dishes, dishes from around the world, with a traditional taste, and a few aspects of innovation, also a handful of recipes featuring cured olives. In the end, I chose dishes that were either enhanced by the olive oil, or dishes that best brought out the flavor of the olive oil, showing both the ingredients and the olive oils to their best advantage. And while many dishes are at their olive-soaked best when lavished with the fragrant, sweet oil, some are more subtly olive-scented, with only a glistening of the oil.

There are nearly endless ways to enjoy the olive and its oil. I offer you a few of my favorites. Enjoy!

Olives are synonymous with the idea of an appetizer before a meal, a few deliciously saline nuggets to munch as you sit in a café watching life unfold on the Cours Selaya, or in the village as you gaze out from your terrace, or, in your kitchen, after a hard day's work. In fact, they are excellent anytime, and anywhere.

Dishes made with olive oil are light and savory, too—like peppers roasted and bathed in the pungent oil; fresh, sweet, raw fennel, drizzled with the lyrical oil; a few grilled fish, brushed with olive oil before they hit the fire. And you should always serve a few chunks of bread to dip into the spicy, flavorful oil that has mingled with the juices of whatever it is cooked with.

Soups, made from vegetables and herbs, are light and vivacious when made with olive oil. This is especially true in the Mediterranean, where such traditional dishes as fish soups, vegetable soups with pistou or pesto, and salad-like, raw-vegetable soups known as "gazpachos" nourish and refresh.

The Simplest Bowl of Olives

Choose a selection of olives in a wide variety; then drain and combine them in a bowl to serve. The delight of big, plump, purple olives next to tiny, cracked, green ones, wrinkled black olives, and round, taupe-colored olives is endless. A few olive leaves add to the visual excitement—if you don't have any, add a few fronds of fresh rosemary, or a fresh bay leaf or two.

Preparation: 10–15 minutes

Olive-Cheese Platter

olive e formaggio

This utterly simple dish is at its best when it has had a chance to macerate (soften by steeping in liquids) a bit. Serve with chunks of bread for dipping, on a hot day, alongside a tomato salad, and perhaps sprinkle a few drops of balsamic or sherry-wine vinegar over the olives.

SERVES 4
- 2 cups or so black olives, such as Italian oil- or salt-cured, or big, fat, salty Greek olives or the little Cretan olives

- ¼ cup/2–3 fl oz extra-virgin olive oil
- 1 tsp chopped, fresh rosemary
- ½ cup feta, mozzarella, or other white cheese, cut into cubes

Preparation: 5 minutes

Marinating time: 1 hour

Combine the olives with the oil, rosemary, and cheese. Leave to marinate for at least an hour.

Vegetable and Olive Oil Feast

pinzimonio

The fresh vegetables are enhanced by the star of the dish, the fruity, flavorful olive oil. Each person places a little olive oil in a saucer, seasons it with a few sprinklings of salt, pepper, and a squeeze of lemon or drop of balsamic vinegar if desired, then dips the fresh vegetables into this simple little sauce.

SERVES 4
- I red bell pepper, cut into strips
- I bulb sweet fennel, cut into strips and tossed lightly with lemon juice to keep it from discoloring
- I Belgian endive, cut into spears
- 2–3 artichoke hearts; raw if young and tender, blanched if they have grown a choke; toss in lemon juice to keep them from discoloring
- ½ cucumber, sliced
- ½–I head raddichio, cored and leaves separated
- 4 ribs of celery, cut into large pieces

- Handful of arugula leaves
- Handful of pea greens
- Sweet young carrots, blanched or briefly steamed
- 4 ripe tomatoes, cut into wedges
- Cruet of good fruity olive oil: allow 1–2 Tbsp per person
- 3–4 Tbsp balsamic vinegar per person
- 3–4 Tbsp flaked or coarse grain sea salt per person
- 1–2 Tbsp coarsely ground black pepper per person
- I lemon, cut into wedges

Preparation: 15–30 minutes

❶ Arrange the vegetables on a platter or in a basket. Diners at the table might want to chop their vegetables into smaller pieces, so be sure to provide sharp knives for such a desire.

❷ Arrange the cruet of olive oil on the table. For each person arrange a sauce of balsamic vinegar, a little plate of coarse salt, black pepper, and lemon wedges. Serve each person a saucer with which to mix their sauce and vegetables.

Olive and Fennel Relish

salsa d'olive

While I like oil-cured olives in this relish, any good, strong, black olive is delicious, though try to choose one that is not too tangy or vinegary. Serve this as an antipasto with crusty bread, or with fire-broiled tuna.

SERVES 4
- I fennel bulb, diced
- Juice of ½ lemon

- 10–15 oil-cured black olives, pitted and cut into quarters or halves
- 2–3 Tbsp extra-virgin olive oil

Preparation: 10 minutes

Combine all of the ingredients and enjoy.

Braised Artichokes

carciofi alla romana

Artichokes, stuffed with mint, parsley, and garlic, then braised with olive oil and lemon, is one of the delights of the Roman kitchen.

SERVES 4–6

- ¼ cup parsley, finely chopped
- 2–3 Tbsp finely chopped, fresh, mint leaves, or more as desired
- 8–10 cloves garlic, finely chopped
- Salt and pepper to taste
- ½ cup extra-virgin olive oil
- 8 medium-sized artichokes
- Juice of 1 lemon

Preparation: 30–40 minutes

Cooking time: 1 hour, 15 minutes

❶ Combine the parsley with the mint, garlic, salt, and pepper, and about three tablespoons of the olive oil, or enough to form a paste. Leave to marinate while you prepare the artichokes.

❷ Remove the hard leaves from the artichokes, and cut away the sharp top from the tender, inner leaves. Pull the center open, and scoop out the thistly inside, using a spoon and a sharp paring knife.

❸ Stuff the inside of each artichoke with the herby mixture, then lay the artichokes in a baking dish in a single layer. Sprinkle with salt, pepper, and any leftover herby mixture, then with the remaining olive oil and lemon juice, adding enough water to cover the artichokes. Cover with a lid or with foil.

❹ Bake in a preheated (350°F) oven for about an hour. Remove the lid to taste the sauce; if it lacks flavor, pour it into a saucepan and reduce until it condenses and intensifies—you want the liquid to be evaporated to isolate the flavorful oil. Once that has happened, stop boiling immediately. Season with salt, pepper, and lemon juice, then pour it back over the artichokes. Eat hot or cold.

Balkan Grated Turnip

with olive oil and walnuts

We sampled this a few years ago on a visit to a wedding in Bulgaria. Although olive oil is not a particularly Bulgarian flavor, guests from Macedonia brought a bottle, and drizzled it onto this turnip salad.

SERVES 4

- 2–3 young, medium-sized turnips, peeled
- 1 Tbsp extra-virgin olive oil, or as desired
- Juice of ½ or more lemon, to taste
- Sea salt to taste
- 2–3 Tbsp walnut pieces
- ½ tsp chopped parsley or a sprig of parsley to garnish

Preparation: 10–15 minutes

Coarsely grate the turnip, then toss it with the olive oil, lemon, and salt. Mound it onto a plate, then garnish with walnuts. Serve sprinkled with parsley or garnished with a parsley sprig.

appetizers and snacks

▲ *Selection of antipasto*

Roasted Tomatoes

pomadori arrosti

Delicious served with crusty bread to soak up the juices and tomato. Capers, olives, or anchovies can be scattered on top, for a salty Mediterranean flavor.

SERVES 4

- 12 small to medium-sized ripe flavorful tomatoes
- 4–5 Tbsp extra-virgin olive oil
- ½–1 tsp balsamic vinegar, or to taste
- Coarse sea salt to taste
- 3–5 garlic cloves, finely chopped
- Handful of fresh basil, coarsely torn

Preparation: 5–10 minutes

Cooking time: 1–1½ hours

❶ Arrange the tomatoes in a casserole, about an inch or so apart, and drizzle with about 1–2 teaspoons of the olive oil.

❷ Preheat the oven to 425°F and place the tomatoes inside to roast for 20 minutes. Reduce the heat to 325°F and continue to cook for a further 30 minutes. The skins should be darkened and cracked. Remove from the oven and let cool, preferably overnight when their juices will thicken.

❸ Remove the skins of the tomatoes and squeeze them of their juices, letting the juices run back onto the tomatoes.

❹ Serve at room temperature, drizzled with olive oil, balsamic vinegar, and sprinkled with salt, garlic, and basil.

Potato and Lentil Spread

brandade aux lentilles

Brandade is a thick paste of potatoes and usually cooked salt cod. Instead of salt cod, vegetables are sometimes used.

SERVES 4

- 4 medium-sized potatoes, peeled and cut into quarters or chunks
- 1 cup cooked, drained, green lentils
- ½ tsp Herbes de Provence to taste
- 4–5 garlic cloves, minced
- Sea salt
- ½–¾ cup extra-virgin olive oil
- Juice of ½ lemon or to taste
- Freshly ground black pepper to taste
- 8–10 Mediterranean black olives
- Handful of arugula or other greens
- Sprinkle of paprika

Preparation: 15 minutes

Cooking time: 15–20 minutes

❶ Boil the potatoes until they are very tender, about 15 minutes, then drain and mash. Meanwhile, grind or purée the lentils and mix with the potatoes. Add the Herbes de Provence to taste.

❷ Crush the garlic with the salt until it forms a paste, then stir it into the vegetables and begin to work in the olive oil, a tablespoon or two at a time, until all of the olive oil is absorbed. If the olive oil begins to be not easily absorbed, don't add any more.

❸ Season with lemon juice, salt, and pepper to taste, and mound it onto a plate. Garnish with olives and arugula leaves, and sprinkle with paprika for color.

Middle-Eastern Breakfast

feta-yogurt spread, olive oil, and dukkah

Bread and olive oil, with a tangy mound of feta-yogurt spread, and a sprinkling of the Middle-Eastern spice-and-nut mixture, called dukkah, *is the classic breakfast throughout the Middle East, and also makes a great mid-afternoon snack, or a good appetizer to prepare with any Middle-Eastern oils if you have access to them, such as Lebanese or Israeli, though it is delicious made with any good, extra-virgin olive oil.*

SERVES 4

- 3 cloves garlic, chopped
- 4–6 oz feta cheese, crumbled
- 3–4 tablespoons yogurt
- Extra-virgin olive oil, as needed
- ½–1 tsp each: ground coriander, cumin, and thyme
- 1–2 tsp sumac
- 2–3 Tbsp each: coarsely ground, toasted sesame seeds, and filberts or almonds

TO GARNISH:

- 10–15 Mediterranean black olives, or as many as you like
- 3–5 scallions
- 4–8 warm pita-bread pockets or 2–4 warm naan breads, torn into pieces and wrapped in a cloth

Preparation: 30 minutes

❶ Combine the chopped garlic with the crumbled feta, yogurt, and a tablespoon or two of olive oil. Mound onto a plate.

❷ Combine the coriander, cumin, thyme, sumac, sesame seeds, and nuts to make the *dukkah*.

❸ Pour a few tablespoons of olive oil onto a saucer, and sprinkle with the *dukkah* mixture. Garnish with the olives and scallions.

❹ Serve the plate of cheese mixture and the olive oil–spice platter with the pita or naan. Let each person dip into the various mixtures as desired, mixing and combining to taste.

Cretan Sandwich

ntakos

This utterly refreshing salad-sandwich, pronounced "Dakos," is courtesy of my friend Panos Manuelides of OdySea imports. This dish hails from the island of Crete, and is healthy, delicious, and rich in vitamins and fiber. It also wonderfully shows off the fresh flavors of the vegetables and the olive oil.

It can be made ahead of time, indeed it must be; fewer things can revive a person on a hot, sultry day as this. It is also perfect for picnics, as I have often seen families sitting in the fields on an afternoon, sharing this dish.

SERVES 4
- 4 *paximadia* or other whole-wheat crackers
- 12 or more ripe, sweet, juicy tomatoes
- Sprinkle of oregano
- Approximately 3 dozen black Greek or other Mediterranean olives, pitted and cut into pieces
- 2 cups/6–8 oz or so *kefalotiri* or pecorino cheese, thinly sliced or shaved
- ¼–½ cup/3–4 fl oz olive oil (preferably Greek), or as needed

Preparation: 10–15 minutes

Marinating time: at least 4 hours

❶ Arrange the *paximadia* on a platter or on plates. Layer with the tomatoes, oregano, olives, and cheese, and drizzle with olive oil.

❷ Leave to marinate for at least four hours at room temperature, then serve.

Italian-style Marinated Carrots

carotte italiane

Carrots, cooked only until just tender and bright-orange in color, have a sweet flavor that is quite enhanced by a good splash of olive oil and vinegar. With a scattering of garlic and parsley, they make a terrific antipasto.

The following recipe was given to me by an elderly Italian farmer and his even-more elderly mother, who loved the dish.

SERVES 4
- 8–10 medium-sized carrots
- 3–5 cloves garlic, chopped
- 3 Tbsp extra-virgin olive oil, or more to taste
- 2 Tbsp raspberry vinegar (or a mild wine vinegar), or more to taste
- 3–5 Tbsp chopped parsley
- Salt and pepper to taste

Preparation: 15 minutes

Cooking time: 15 minutes

Cut the carrots into about ¼-inch-thick slices. Steam or boil them until they are tender, about 15 minutes. Drain well, then toss with the remaining ingredients. If the carrots are not very sweet, add a tiny pinch of sugar to the cooking water.

▲ Cretan Sandwich

33

Tomato and Basil Bruschetta

bruschetta al pomodoro

One summer, we rented a little house outside of Florence, and noticed that bruschetta al pomodoro was on the menu everywhere. This recipe was brought to me from Tuscany, where it is also known as fettunta, by the Mediterranean-based writer Paul Richardson.

SERVES 4
- 4 thick slices of Italian country bread (or French bread, if not available)
- ½ cup olive oil, preferably Tuscan (or more, as needed)
- 6 very ripe, flavorful tomatoes, diced
- Handful of fresh, sweet basil leaves, torn
- 4 garlic cloves, chopped
- Coarse sea salt, to taste

Preparation: 10–15 minutes

Cooking time: 10–15 minutes

❶ Brush the bread with several tablespoons of the olive oil, then toast it on a baking sheet at 425°F for about 15 minutes, turning once or twice or until crisp and golden brown.

❷ Combine the tomatoes with the rest of the olive oil, basil, and garlic, sprinkle with coarse sea salt, and serve.

Mediterranean Fried Squid

Serve crispy-fried squid with a plate of olives, wedges of lemon, and perhaps slices of tomatoes topped with feta cubes for a Greek flavor.

SERVES 4–6
- 2½ lbs squid—the smallest you can find
- 1 tsp sea salt
- 2 cups (approximately) all-purpose flour
- 1 cup beer, sparkling wine, or sparkling water
- 2 cups pure olive oil
- ½ cup extra-virgin olive oil

TO SERVE:
- 2 lemons, cut into wedges

Preparation: 15 minutes

Cooking time: 15 minutes

❶ Clean the squid, or have them cleaned for you, then cut them into bite-sized pieces or thin rounds.

❷ Combine the salt with the flour, and place it on a plate, then pour the beer or other soaking liquid into a saucer. Heat the oils together until they are hot, but not smoking; then, working in several batches at a time, dip the squid first in the flour and salt, then quickly into the liquid, then into the hot oil.

❸ Fry over medium heat until they are crisp and lightly browned, five to eight minutes. Remove and place on paper towels to drain, then serve with lemon wedges.

Greek Fire-roasted Eggplant

melanzanosalata

Roasting eggplants over the fire gives them a smoky scent and flavor. You don't need a grill, however, as you can cook them over your stove if you have a gas oven.

Similar dishes are prepared throughout the Mediterranean, from the tahini-enriched baba ghanoush of Lebanon, the creamed-mayonnaise hatzilim salata of Israel, or the tomato-rich French papeton. Olive oil is the secret to all of them, and especially this version: if you have no lemon or garlic, this is still a fine dish, as the flavor of smoky eggplant and olive oil is the essence of the Mediterranean.

SERVES 4

- 3 long, medium-sized eggplants, whole
- Sea salt to taste
- ½ cup extra-virgin olive oil, or as desired
- 1–2 garlic cloves, chopped fine
- Juice of about ½ lemon, or to taste

Preparation: 10–15 minutes

Cooking time: 15–20 minutes

❶ Place the eggplants directly over a flame, either on a grill or on the top of a gas stove; alternatively, in the broiler. Cook slowly, turning and turning until the skin of the eggplants is black and charred, and the flesh is tender. This should take 15–20 minutes.

❷ Remove the eggplants from the heat, and place them in a bowl with a tight-fitting lid. Leave until they cool, for about two hours. Remove from the bowl, taking care to save the smoky liquid that has oozed out. Scoop the flesh from the charred skin, and discard the skin, adding the flesh to the eggplant's smoky juices.

❸ Coarsely chop about two-thirds of the eggplant mixture, and season it with salt. Blend the remaining third of the eggplant purée in the blender with the olive oil, then add it to the rest of the eggplant.

❹ Season with garlic, lemon juice, and salt to taste. Chill until ready to serve.

Mexican Marinated Fish

ceviche

SERVES 4
- 3 cups/12 oz fish fillets, skinned, boned, and cut into bite-sized chunks
- 2 cups/8 oz uncooked shrimp, shelled, cleaned, and deveined
- Juice of 1 orange
- Juice of 5 limes
- 2 very ripe flavorful tomatoes, diced
- 6 scallions or 1 red onion, sliced
- ¼ tsp each, or to taste: sea salt, freshly ground black pepper, oregano leaves
- Pinch of sugar, if needed, to balance the acidity of the ingredients and bring out the flavor of the tomatoes
- 1 fresh, green chile, thinly sliced or chopped
- ⅛–¼ tsp of cumin seeds, or to taste
- 15 or so pimiento-stuffed green olives, sliced or halved
- 4–5 Tbsp coarsely chopped, fresh cilantro
- 1–2 Tbsp extra-virgin olive oil, or to taste

TO SERVE:
- 2 cups shredded lettuce
- Corn tortillas or tortilla chips

Preparation: 20 minutes

Marinating time: overnight

❶ Combine the fish and shrimp with the freshly squeezed orange and lime juice, and leave in a covered, nonreactive bowl (such as glass) to marinate overnight in the refrigerator. Turn the fish and shrimp over once or twice.

❷ Add the diced tomatoes, scallions, salt, pepper, oregano,

Throughout Latin America, fish are marinated in lime and cilantro, seasoned with hot peppers, and drizzled with olive oil. One of my favorite dishes is from Peru: the fish are not raw, but are tiny, sardine-like creatures fried in olive oil to a crisp, then marinated with lime, carrots, and other vegetables.

You can either use only fish or shrimp in the following recipe. The combination of olives and olive oil with the citrus-flavored fish and shrimp is particularly good. You can garnish it with avocado slices or add fire with more chiles, if you like—whichever way you choose, ceviche is a refreshing, invigorating way to begin a warm-weather feast.

sugar (if needed), green chile, cumin seeds, green olives, cilantro, and olive oil. Chill until ready to serve. If it seems too wet, drain some of the liquid before serving. Taste for seasoning, and serve garnished with shredded lettuce, and accompanied by warm, soft, corn tortillas or crisp tortilla chips.

Tomato and Pepper Gazpacho

salmorejo

SERVES 4

- 4 oz stale, country bread
- ¾ cup extra-virgin olive oil, Andalusian, preferably
- 1½ cups/12 oz ripe tomatoes, diced
- 1 green bell pepper, diced
- 4 garlic cloves, finely chopped
- 2 Tbsp sherry vinegar
- Salt to taste

TO GARNISH:

- 1 hard-cooked egg, chopped
- 1 slice of country bread, cut into cubes and browned in olive oil
- 1–2 slices Serrano or parma ham, cut into thin strips

Preparation: 15 minutes

Chilling time: 1 hour

---

❶ Cut or break the bread into bite-sized pieces, and place in the food processor or blender. Pour cold water to cover over it, leave for a moment or two, then drain.

❷ Add the olive oil, tomatoes, green bell pepper, garlic, sherry vinegar, and salt, and blend until a thick purée forms. Season to taste, and chill until ready to serve with the optional garnishes.

This thick gazpacho gets its creaminess from puréed, stale bread and olive oil, blended with the vegetables. It is a specialty of Andalusia, traditionally made by pounding, though a blender works quite nicely. Salmorejo may be served as a starter, or as a tapa in tiny cups for sipping.

A crisp, crunchy garnish is a nice addition, since the soup is so thick and creamy—diced, hard-cooked egg, crisp little croutons, and thin shreds of Spanish ham are traditionally used. I sometimes add a few shreds of fresh mint leaves.

appetizers and snacks

37

Green "pistou" Soup

soupe de légumes vertes au pistou

The lovely, all-green color is an unexpected delight. Add pasta if you like—to make a classic pistou add a handful of cooked vermicelli, broken and squiggling into the pot.

SERVES 4
- 1 leek, chopped
- 2 onions, chopped
- 5 garlic cloves, coarsely chopped
- 3 Tbsp extra-virgin olive oil
- 3 medium-sized ripe, yellow tomatoes, diced
- Salt and pepper to taste
- A pinch of sugar, to balance out the acidity of the tomatoes
- 1¾ pints vegetable broth
- 1¾ pints water
- 3 small zucchini, cut into bite-sized pieces
- 1 medium-sized, waxy potato, peeled and diced
- ¼ cabbage, thinly sliced
- ¾ cup fresh, white, shell beans, such as coco, shelled and cooked, or cooked cannellini
- 8 chard or spinach leaves, thinly sliced
- ¼–½ bunch broccoli, cut into flowerets
- Handful of green beans, cut into bite-sized pieces
- 1 batch of pistou (pg. 118)
- Extra grated Parmesan, to taste

Preparation: 15–20 minutes

Cooking time: 20–30 minutes

❶ Lightly sauté the leek, onion, and garlic in the olive oil until they are soft, then add the tomatoes and cook over a medium heat for about 10 minutes. Add the vegetable broth and water, the zucchinis, potato, and cabbage, and continue to cook over a medium heat until the potatoes are just tender and the zucchinis are quite soft. The cabbage will be soft by now, too.

❷ Add the white beans, chard or spinach leaves, broccoli, and green beans, and cook for about another five minutes, or until the broccoli and green beans are cooked through.

❸ Serve immediately in bowls with a tablespoon or two of pistou stirred in, accompanied by the grated Parmesan, to taste. Do not heat the pistou, or its fragrance will dissipate.

▲ Green "pistou" Soup

Fish Soup

soupe de poisson

Classic fare from the South of France, this is only one of the many soups prepared. From Italy to Greece, Spain, France, North Africa, Turkey, and further points around the Mediterranean, fish stews and soups are prepared by sautéeing onions and garlic in olive oil, then adding a bit of tomato and whatever fish are available. It is the flavor of the olive oil simmering with the fish that so captures the essential Mediterranean flavor.

SERVES 6
- 1 onion, chopped
- 1 leek, chopped
- 5 cloves garlic, crushed with a pinch or two of salt
- $\frac{1}{4}$–$\frac{1}{2}$ cup extra-virgin olive oil, or as needed
- $2\frac{1}{4}$ lbs mixed fish/fish trimmings, cut into small pieces
- 1 cup ripe tomatoes, chopped
- 2 bay leaves
- Pinch each of thyme, fennel seeds, and chopped rosemary
- $\frac{1}{8}$ tsp grated orange rind
- 1 tsp chopped parsley
- 1 cup dry white wine
- Enough water to cover the fish
- 1 cup prepared fish, vegetable, or chicken broth
- $\frac{1}{2}$ tsp saffron
- Sea salt to taste
- Several pinches of cayenne pepper
- 4 slices of stale bread, cut into slices and lightly toasted, or baked in the oven until crisp
- 1 clove garlic, cut into halves
- Freshly grated cheese (I recommend a combination of Gruyère and Parmesan)
- Garlic-and-Chili Mayo (pg. 120)

Preparation: 20–25 minutes

Cooking time: 1–1$\frac{1}{2}$ hours

❶ Lightly sauté the chopped onion and leek, and crushed garlic in a few tablespoons of the olive oil, then add the fish, and lightly sauté together, adding more olive oil if needed.

❷ Add the chopped tomatoes, bay leaves, thyme, fennel, rosemary, grated orange rind, parsley, wine, water, and broth, and bring to a boil. Lower the heat, then simmer until the fish are soft, about 40 minutes.

❸ Add the saffron, salt, and cayenne pepper, then bring the soup to a boil once again, and cook on high heat for about 20 minutes, so that all of the fish nutrients permeate the soup.

❹ Meanwhile, rub the garlic onto both sides of the bread and drizzle with olive oil.

❺ Strain the soup, pressing hard to extract all of the flavor. Pour the soup into the pan and reheat. Spread the bread with the mayo, sprinkle with cheese, and serve, floating the mayo-topped toasts on top of the soup.

appetizers and snacks

A raw vegetable is a raw vegetable, but a raw vegetable with a splash of olive oil is a salad. Most salads are at their crispest, most-flavorful best when dressed with olive oil. Even many nut oils, such as walnut or filbert, are delicious when combined with olive oil.

Balsamic vinegar, red-wine vinegar, lemon juice, even a dash of sour cream are all lovely, tart additions to your salad, and crusty bread for sopping up the dressing is a must.

salads

Spinach and Goat's Cheese Salad

salade aux épinards et chèvre

Such a simple salad that it makes you want to smile from the sheer pleasure of its freshness and taste. I sometimes add a few leaves of mint, shredded, to the spinach and scallions.

SERVES 4

- 1 bunch of young spinach leaves
- 2 scallions, thinly sliced
- 1 slice goat's cheese (the kind with the rind), cut into slices or bite-sized pieces
- 1–2 garlic cloves, chopped fine
- 2–3 Tbsp extra-virgin olive oil (Provençal variety is recommended)
- ½ tsp balsamic vinegar, or to taste
- Salt and pepper, as desired

Preparation: 10 minutes

❶ Roll the spinach leaves carefully, then slice them very thinly with a sharp knife, cutting them nearly into shreds. The thinner they are, the lighter and fluffier they will be on the plate. Toss the spinach with the scallions, and arrange on a plate.

❷ Top the spinach mixture with the goat's cheese, and sprinkle with the garlic, then drizzle olive oil and balsamic vinegar over the top, seasoning with salt and pepper as desired.

Partridges from my Garden

"perdices de mi huerto"

Wedges of little gem or miniature lettuces, chilled and arranged in a spoke-like circle, sprinkled with olive oil, lemon juice, and grains of coarse salt, is a Murcian specialty from Spain's southeastern coast. Though untraditional, balsamic vinegar is delicious used in place of the lemon.

The dish, made for me by British–Spanish writer Paul Richardson, is particularly good when served with a hefty main course such as pasta, on a hot and sultry day. I recently prepared this using the Cretan Messara olive oil, and it was superb: the oil imparted an earthy-but-clean olive flavor, and managed not to overwhelm the subtle flavor of the little lettuces.

SERVES 4

- 4 little gem lettuces, cut into wedges lengthwise, and chilled
- Juice of 1–2 lemons, as desired
- 3 Tbsp extra-virgin olive oil
- Sprinkling of coarse salt

Preparation: 5 minutes

Arrange the lettuces on a serving plate in spoke-like manner, then sprinkle with lemon, olive oil, and coarse salt. This dish is best served and eaten immediately.

▲ *Spinach and Goat's Cheese Salad*

Andalusian Chopped-Vegetable

salad with cumin vinaigrette

When the weather is hot, I make up vats of this stuff and eat it chilled for any and every meal—even breakfast, when it is exquisitely refreshing. It makes a very nice relish for a bocadillo, a crusty Spanish sandwich, or for grilled fish.

SERVES 4

- 1 large cucumber, diced
- 3–5 small, ripe tomatoes, diced
- 1 carrot, diced
- 1 red bell pepper, diced
- 1 green bell pepper, diced (add a yellow or orange bell pepper here, too, if desired)
- 3–5 scallions, thinly sliced, or 1 small, chopped onion

- 3–5 garlic cloves, crushed
- ¼ tsp ground cumin or cumin seeds
- Salt to taste
- Juice of 1 lemon
- 1 tsp sherry vinegar or white-wine vinegar
- 3 Tbsp extra-virgin olive oil or to taste (use one with big flavor and aroma)

Preparation: 15 minutes

Combine the cucumber, tomatoes, carrot, red and green bell peppers, scallions, and garlic. Toss with cumin, salt, lemon, sherry or white-wine vinegar, olive oil, and herbs. Taste for seasoning, and chill until ready to eat.

Egyptian-inspired Brown-Bean

and raw vegetable salad

When we docked at dawn in Port Said, Egypt's tiny boats of friendly fishermen ushered us through the harbor. I had only one thing on my mind: "ful" beans, dried fava beans, the meaty, brown bean that is Egypt's national dish, served drizzled with olive oil and a salad garnish. In the following recipe, it is used as a salad; I like to accompany it with big, round, Egyptian flatbreads, but ordinary pita can be substituted.

SERVES 4

- 1½ cups cooked borlotti or ful beans
- 2–3 hard-cooked eggs, preferably still warm, peeled, and diced
- Extra-virgin olive oil, as desired, preferably a North African, Greek, or Turkish oil
- 3–4 garlic cloves, minced
- A large pinch of salt
- 1 onion, chopped
- Handful of arugula, coarsely chopped

- 2–3 ripe tomatoes, chopped
- 1 Tbsp each: cilantro, dill, mint
- 2 lemons, cut into wedges

Preparation: 10 minutes

Cooking time: 5 minutes

❶ Warm the beans in their juices, then drain and arrange on a platter. Garnish with the eggs.

❷ Work several tablespoons of the olive oil into the garlic, then pour this over the beans. Sprinkle the onion, arugula, tomatoes, cilantro, dill, and mint around the top, and garnish with lemon. Drizzle extra olive oil over the top, and serve with a cruet of olive oil and a bowl of coarse sea salt for sprinkling.

▲ *Andalusian Chopped-Vegetable Salad with Cumin Vinaigrette*

Cypriot Village Salad

horiatiki salata

Unlike the village salads I've eaten in Greece, a Cypriot salad contains many more greens—caper stems (be careful, though—these can have thorns); wild seaweed, lightly pickled; rocca (arugula); shredded cabbage (white, green, or red); purslane; cilantro; and pea greens. You never know what greens will appear in your evening salad, as the Cypriots are avid, daily hunters of wild herbs and salad leaves, and all of these go into the foods. The only constant is the glistening, rich, olive oil and the lemon halves plunked onto the table, to squeeze on as you like.

SERVES **4** AS A MAIN COURSE
OR **4–6** AS A STARTER/MEZE

- ½–1 small white cabbage, shredded
- Handful of arugula leaves, chopped
- 3–4 Tbsp chopped, fresh cilantro
- 1–2 Tbsp chopped, fresh parsley
- Wild greens as desired (see above)
- 1 cucumber, diced
- 1 bunch scallions, thinly sliced
- Mint leaves, thinly sliced (optional)
- Extra-virgin olive oil, as desired
- Juice of 1 lemon, plus extra lemon halves, to serve
- Salt and pepper to taste
- 5–6 ripe, juicy tomatoes, quartered
- 10 or more black olives
- 10 or more slightly bitter, green olives
- 1 cup/4 oz feta cheese, cut into slices

Preparation: 10–15 minutes

❶ Combine the cabbage, arugula, cilantro, parsley, wild greens (if using), cucumber, scallions, and mint. Toss with olive oil, lemon, salt, and pepper, then arrange in a bowl.

❷ Top with tomato wedges, olives, and feta, and serve with more olive oil and lemon.

salads

Shepherd's Salad

salade du berger

The bread, cheese, and olive oil in the dressing give this salad the rustic character of southern France.

SERVES **4**

- 12 thin slices of stale baguette (French bread)
- 2 small, round sheep's or goat's cheeses
- Mesclun salad greens (a continental mix, with arugula, field salad, frisée, etc; use other fresh salad greens, if unavailable)
- 3 Tbsp extra-virgin olive oil, preferably French or Ligurian
- 1 Tbsp red-wine vinegar
- 2 slices prosciutto or Parma ham, very thinly sliced into strips
- 1 tsp chopped chervil
- 1 tsp chives
- 1 tsp parsley or fresh tarragon

Preparation: 10–15 minutes

Cooking time: 15–20 minutes

❶ Preheat the oven to 375°F. Toast the slices of bread in the oven until lightly golden and crisp on each side, for about 10 minutes or so.

❷ If using the small, round sheep's or goat's cheeses, cut them into chunks and place a chunk on each piece of toasted baguette. Place on a baking sheet. Bake for 5–8 minutes, or until the cheese warms and just begins to melt.

❸ Meanwhile, toss the mesclun salad greens with the extra-virgin olive oil and red-wine vinegar. Arrange the salad on a platter and place the baguette slices on top.

❹ Sprinkle with the thinly sliced prosciutto or Parma ham and the chopped chervil, chives, parsley and/or tarragon, and serve immediately.

salads

47

Lebanese Salad

Fattoush— tomato, mint, parsley, cucumber, and pita bread salad

This salad of stale bread and salad vegetables from Lebanon, characteristically dressed with lots of olive oil and lemon, is very refreshing. A sprinkle of sumac—a tart, red berry related to poison sumac, but not harmful—gives a distinctive tang. I like to serve it with a helping of yogurt, or yogurt mixed with feta cheese.

SERVES **4**

- 1 large or 2 small cucumbers, diced
- 3 ripe tomatoes, diced
- 1 green bell pepper, diced
- 1 oz (about 8 Tbsp) each: fresh mint, fresh, chopped cilantro, fresh, chopped parsley
- 3 scallions, thinly sliced
- 1 tsp salt
- 3 garlic cloves, chopped

- 3 fl oz extra-virgin olive oil
- Juice of 3 lemons
- 1 tsp sumac
- 3–4 pita breads, stale and lightly toasted, then broken into pieces

Preparation: 15–20 minutes

Chilling time: 1 hour

Combine the cucumbers, tomatoes, bell pepper, mint, cilantro, parsley, scallions, salt, garlic, olive oil, lemon juice, and sumac. Chill for at least one hour. Just before serving, toss with the broken pita breads.

salads

Greek Black-eyed–Pea Salad

with caper dressing

During lunch in a monastery high in the hillside of Mount Hymettus, for the blessing of the olive oil, we were served a vegetarian lunch, which included this delicious salad, accompanied by sweet, ripe, seasonal tomatoes and chunks of crusty bread.

SERVES 4, AS A MEZE OR SIDE DISH

- 2 cups raw, black-eyed peas
- 2–3 Tbsp red-wine vinegar, or to taste
- Pinch of sugar
- 2–3 Tbsp capers
- 3 fl oz extra-virgin olive oil, preferably a Greek or Cypriot oil, or to taste
- 3–5 garlic cloves, chopped
- Large pinch of oregano leaves, crushed between the fingers
- 2–3 Tbsp fresh, chopped parsley, preferably the flat-leaf variety
- ¼ onion, red or white, chopped
- Salt and pepper to taste
- Mixed bed of salad greens (young cabbage, arugula, etc.)

Preparation: 15 minutes

Cooking time: 40–60 minutes

❶ Place the black-eyed peas in a saucepan and cover with cold water. Bring to a boil, then reduce the heat and simmer over a low heat until the beans are tender, for approximately 40 minutes. Pre-soaking for at least an hour, or even better overnight, will reduce the cooking time, resulting in more-tender peas.

❷ Drain the peas and toss them with the vinegar, sugar, capers, olive oil, garlic, oregano, parsley, onion, salt, and pepper. Serve on the bed of mixed salad greens and eat right away.

salads

Roasted Beet-and-Potato Salad

salade aux betteraves

If you have the opportunity to obtain yellow beets in addition to red beets, this adds an extra dimension of color, plus it's very colorful to find a scarlet vegetable sitting in yellow chunks on the plate, as it looks like a delightfully edible joke! If yellow beets are not available, you may substitute yellow tomatoes, or bell peppers, to achieve the same effect. Also yellow, Finnish, white Cypriot, or steely blue potatoes each add their rainbow hues. All of these colors topped with a sprinkling of fresh dill, cilantro, and scallion, make a beautiful, jewel-like salad, as ravishing to eat as it is to look at.

SERVES 4

- 1 lb raw, yellow beets
- 1 lb raw, red beets (if unavailable, use already-cooked beets, and omit the step of roasting the vegetable)
- 1 tsp sugar
- ½ cup extra-virgin olive oil
- 1 lb small potatoes, preferably a mixture of colors: pink fir, blue, yellow Finnish, etc., or use baby or new potatoes, such as Cypriot
- 1 cup yellow tomatoes
- 1 cup cherry tomatoes, halved
- Salt and cayenne pepper to taste
- Juice of 1–2 lemons
- ½ tsp powdered cumin (or more to taste)
- 2 Tbsp chopped, fresh cilantro
- 2 Tbsp chopped, fresh dill
- 2 scallions, thinly sliced
- 8–10 purple-black olives (wrinkled salt- or oil-cured olives; Gaeta are recommended)

Preparation: 20 minutes

Cooking time: 1 hour

❶ Place the beets in a roasting pan in one layer, but preferably not touching each other. If you are using both yellow and scarlet beets, roast them in separate pans. Turn the oven up to 375°F and roast the beets for about an hour, or until they are tender; larger beets will take longer; smaller, more tender ones take less time. Alternatively, you may steam or boil them, but the flavor is richest when roasted.

❷ When they are cool enough to handle, slip the skins off and cube them. Toss them with the sugar, and about a third of the lemon juice and olive oil; then set aside (if you have both yellow and scarlet beets, be sure to dress them separately).

❸ Boil or steam the potatoes until they are just cooked, then remove them from the heat and leave them to cool slightly. When they are cool enough to handle, slip their skins off and cube them. Dress with salt and cayenne pepper, and about a third of the fresh lemon juice and olive oil; then set aside.

❹ Arrange the beets, potatoes, and tomatoes on the plate, and sprinkle with cumin. Dress with the remaining freshly squeezed lemon juice and olive oil, and scatter the chopped dill, cilantro, scallions, and olives over the mixture.

salads

Olive oil is a perfect companion to grilled foods, whether used in marinades or sauces, or simply brushed onto whatever meets the heat of the grill.

If you do nothing else, take your cruet of olive oil and a few wedges of lemon to the grill. Brush whatever you are cooking with the oil, then cook it over the coals. When tender, remove to a plate and drizzle with more olive oil; then squirt lemon over it, and sprinkle with whatever herbs you desire. Voilà!—the essential Mediterranean grill.

Grilled Halloumi

with olive oil, brandy, and lemon-herb-marinade

Grilled halloumi cheese is one of the great and simple pleasures of eating in Cyprus. When cold, halloumi is rubbery, slightly tough, and quite salty—when heated over a grill, it grows soft and supple, imbued with the deliciously smoky scent of the fire.

SERVES 4

• Approximately 2–2¼ cups halloumi cheese, cut into ¼-inch slices
• 2 Tbsp extra-virgin olive oil, preferably a Greek or Cypriot variety
• ⅛ cup brandy
• Several pinches of Herbes de Provence or Bouquet Garni
• 1 garlic clove, chopped
• 1 lemon, cut into halves

Preparation: 5 minutes
Marinating time: 1½–2 hours
Cooking time: 5 minutes

❶ Combine the cheese with the olive oil, brandy, herbs, garlic, and the juice of one lemon half. Leave to marinate for 30 minutes to several hours.

❷ Cut the remaining lemon into wedges, and set aside.

❸ Cook the halloumi cheese quickly on a hot grill, letting it brown lightly on each side, but be careful that it doesn't melt through the grate. You can also use a pan to prevent this.

❹ Remove from the hot fire, and serve immediately, garnished with the remaining lemon wedges.

Pork or Lamb Souvlaki

Tangy, garlicky, and richly scented with olive oil, this dish is inspired by souvlaki I ate in a taverna in Crete 20 years ago.

SERVES 6

• 2½ lbs boneless pork or lamb, either leg or shoulder; use a cut with lean, tender meat and a bit of fat at the edge, cut into bite-sized pieces
• 1 onion, chopped fine
• 5 garlic cloves, minced
• ½ cup/3–4 fl oz lemon juice

Preparation: 30 minutes
Marinating time: at least 2 hours
Cooking time: 10 minutes

• ½ cup/3–4 fl oz extra-virgin olive oil
• 1 tsp oregano, crushed
• Salt and black pepper
TO SERVE:
• Chopped parsley, olives, pita bread, lemon wedges for pork, or unsweetened yogurt for lamb

❶ Combine the pork or lamb with the onion, garlic, lemon juice, olive oil, oregano, salt, and pepper. Leave to marinate for at least three hours at room

temperature, or up to two days in the refrigerator.

❷ Remove from the marinade, and skewer onto soaked bamboo or metal skewers. Grill over medium-hot coals for about eight minutes, turning occasionally. Serve immediately with parsley, olives, and lemon, resting on a piece of pita bread. If using lamb, serve a bowl of unsweetened yogurt on the side.

▲ *Grilled Halloumi*

Grilled Vegetables

les légumes grillés

I ate these at the feast of St. Laurent, the patron saint of grilling, who, appropriately enough, was martyred by being burnt at the stake.

SERVES 4

- 3 zucchini, cut into ¼-in slices
- 1 red bell pepper, cut into wide wedges or halves
- 1 yellow bell pepper, cut into quarters
- 1 green bell pepper, cut into quarters
- 3 Tbsp extra-virgin olive oil
- 5 garlic cloves, chopped
- Juice of ½ lemon or 1 Tbsp balsamic or white-wine vinegar
- 14–20 cherry tomatoes
- Bamboo skewers, soaked in cold water for 30 minutes, or metal skewers
- Salt and pepper
- 2–3 Tbsp pesto or fresh basil, chopped fine and puréed with a little garlic and olive oil

Preparation: 10–15 minutes

Marinating time: 30 minutes

Cooking time: 10–15 minutes

❶ Combine the zucchini and peppers with the olive oil, garlic, and lemon or vinegar. Marinate for about 30 minutes if possible.

❷ Meanwhile, thread the cherry tomatoes onto the skewers. Remove the zucchini and peppers from the marinade, saving the marinade to dress the vegetables afterward.

❸ Grill the vegetables over a medium-high heat until they are lightly charred and browned in places, and tender all the way through. Remove from the grill, and return to the marinade. Let stand until vegetables are easy enough to handle.

❹ Meanwhile, grill the cherry tomatoes for about 5 minutes on each side or until slightly browned. Remove from the grill.

❺ Dice the zucchini and peppers, and slice the tomatoes. Season, combine the juices with the pesto, pour over the vegetables, and serve, dressed in a little extra olive oil if desired.

Fire-Grilled Fish

I like this dish with a simple salad of peppers, sliced fennel, and black Niçoise olives, awash in olive oil and lemon. Any small fish are delicious served in this simple style.

SERVES 4

- 2½ lbs small fish, cleaned of their insides but with heads and tails intact
- ½ cup/4–5 fl oz extra-virgin olive oil, Tuscan or Provençal, preferably
- Salt, pepper, and oregano to taste
- 2–3 garlic cloves, minced
- 1 Tbsp fresh chopped parsley
- Juice of 1 lemon

Preparation: 10 minutes

Cooking time: 5–10 minutes

❶ Brush the fish with olive oil, salt, pepper, and oregano, then cook them over an open fire for about three minutes on each side. Remove from the grill.

❷ Combine the garlic with the parsley and lemon, then stir in as much of the olive oil as you can. Arrange the fish on a serving plate and pour over the seasoned olive oil. Serve immediately.

▲ Grilled Vegetables

Middle-Eastern Style Swordfish

smak meschwi bi tahini

SERVES 4

- 2½ lbs swordfish
- 1 onion, grated
- 8 garlic cloves, chopped
- Juice of 2 lemons
- ½ cup extra-virgin olive oil
- Several bay leaves
- Salt and pepper

TAHINI SAUCE

- ¾ cup/6 oz sesame paste (tahini)
- 2 garlic cloves, chopped
- Few dashes of hot-pepper sauce
- Few pinches of cumin
- Salt and pepper
- Juice of 1 lemon
- 2 Tbsp extra-virgin olive oil
- ½ cup water, or enough to make a smooth, thick sauce
- Few sprigs fresh oregano
- Lemon wedges to garnish

Preparation: 20 minutes

Marinating time: at least 1 hour

Cooking time: 8–10 minutes

Eaten throughout the Mediterranean, swordfish is usually marinated with bay leaves, lemon, and lots of olive oil, then skewered and cooked over a wood fire. In Turkey, you'll find it dressed with a walnut sauce; in the eastern Mediterranean, the sauce will likely be a classic tahini. I like a few wedges of lemon, and a dab or two of harissa on the side.

❶ Combine the fish, onion, garlic, lemon juice, olive oil, bay leaves, salt, and pepper. Marinate for at least an hour, preferably overnight in the refrigerator.

❷ Skewer the marinated fish and bay leaves on either soaked bamboo (30 minutes in cold water) or metal skewers, alternating the fish cubes with the bay leaves. Though you don't eat the bay leaves, they perfume the fish delightfully.

❸ Grill over a medium–low charcoal fire for about 8 minutes, turning to cook evenly.

❹ Meanwhile, mix the tahini with the garlic, hot-pepper sauce, cumin, salt, pepper, lemon juice, and olive oil, then slowly stir in the water until it has the desired consistency. Taste for seasoning.

❺ Serve with fresh oregano, accompanied by lemon wedges and the tahini sauce on the side.

Middle-Eastern Grilled Eggplant

with pomegranate-and-mint salad

This dish of thinly sliced, grilled eggplant is served with a sweet-and-sour pomegranate dressing, a leafy green salad, and fresh, fragrant mint. It is an excellent accompaniment to grilled chicken or lamb.

SERVES **4**

- 1 eggplant
- Extra-virgin olive oil, as needed
- 5 garlic cloves, chopped
- 4–5 Tbsp pomegranate syrup or molasses (available in Middle-Eastern delis)
- 6 Tbsp balsamic vinegar
- Handful of mixed salad leaves
- Salt and pepper to taste
- 3 Tbsp mint leaves, chopped fine

Preparation: 20 minutes

Cooking time: 20 minutes

❶ Slice the eggplant and brush slices with olive oil.

❷ Grill the eggplant slices until they are lightly browned and tender inside. Remove from the grill and rub them with the garlic.

❸ Combine the pomegranate syrup with an equal amount of the balsamic vinegar and set aside, then toss the greens with a little olive oil and the remaining balsamic vinegar to taste, adding salt and pepper as desired.

❹ Serve the eggplant slices sprinkled with the pomegranate-and-vinegar syrup and chopped mint, and the mixed-salad leaves on the side.

Roasted Artichokes

anginares psites sta karvouna

Throughout the sun-baked Mediterranean, artichokes are bathed in olive oil and cooked over an open fire. In Sicily, they are actually set directly into the hot coals after the rest of the meal has cooked, and the coals are smoldering; artichokes cooked this way are sublime.

If the artichokes are slightly tough and a little bitter, as they often are, blanch them first and scoop out their sharp, thistly insides. If the artichokes you are using are tender and sweet, with no choke, then you don't need to blanch them or scoop out their insides. So, know your artichokes before you proceed with the recipe. (A good way to check for bitterness is to taste a little bit first when raw. As for the thistle and choke, you can easily see and feel if it needs to be removed.)

On the Greek islands of Kythera and Zakynthos, you still sometimes find artichokes sold as street food, smoky and tender, its leaves to be dunked into more olive oil, lemon juice, and coarse sea salt.

SERVES 4
- 4 medium–large-sized artichokes
- 8 garlic cloves, sliced
- 3 Tbsp lemon juice
- 3 fl oz extra-virgin olive oil, either Greek or Apulian
- Salt and pepper to taste
- 2–3 Tbsp mixed herbs, such as parsley, oregano, marjoram, or thyme
- Coarse salt or aioli (pg. 120) to serve

Preparation: 1 hour

Cooking time: 40 minutes

❶ If necessary, blanch the artichokes until they are part-tender, about 15 minutes, then remove and drain. Cut the artichokes in half and remove the fuzzy, inside choke and thistly middle. Place them in a dish, and marinate with the garlic, lemon juice, olive oil, salt, pepper, and herbs for at least an hour. If not blanching or scooping out the insides, simply pull the leaves apart a bit, then pour over the marinade so that it falls into the crevices of the leaves.

❷ Remove the artichokes from the marinade, reserving the marinade, and grill them on each side, until they are lightly browned, basting continually with the reserved marinade. For blanched artichokes, you will only need to grill them for 10 to 15 minutes; for raw ones, allow about 40 minutes, or until the leaves pull off easily.

❸ Serve with more olive oil, lemon juice, and coarse salt, or with a bowl of aioli.

hot off the grill

Olive oil and vegetables are lusty companions
whether you are serving a side dish or a main
course. Almost any vegetable is improved by a
hefty splashing of this smooth, sweet oil. The
finest of the season's vegetables are picked from
the garden, steamed or boiled, then served awash
with olive oil and lemon. In Greek tavernas, the
most-common recipe you will find is whatever
vegetable is ripe in the garden or market, stewed
with onions, garlic, perhaps tomatoes, and
lashings of olive oil. The oil takes on the flavor
of the vegetables, and makes the most luscious
sauce for dipping bread imaginable.

vegetable dishes

Savory Roasted Pumpkin

Drizzling the pumpkin or squash with olive oil and balsamic vinegar, then baking it, is a simple, healthy way to serve this rich, sweet vegetable.

SERVES 4
- 2½ lbs pumpkin, or large orange squash or other winter squash, cut into several pieces
- 3–5 garlic cloves, chopped
- 3 Tbsp extra-virgin olive oil
- 1 tsp balsamic vinegar
- Sea salt and freshly ground black pepper, to taste
- A pinch of mild, red chili powder, to taste
- A pinch of oregano or sage, to taste

Preparation: 5–10 minutes

Cooking time: 1 hour

❶ Arrange the squash or pumpkin on a baking sheet, and sprinkle all of the other ingredients over it.

❷ Cover tightly with foil, and bake for about an hour at 350°F, or until the pumpkin or squash is tender. Unwrap and serve hot.

VARIATION

Leftovers may be mashed with a splash of extra-virgin olive oil, a little chopped garlic, a few drops of lemon, and served as a antipasto or meze-type salad.

vegetable dishes

66

Lentils with Chorizo

lentejs con chorizo

I like to make this delectable lentil dish with French puy lentils—they stay together when cooked, and never go mushy. This dish is surprisingly easy to toss together, and robustly delicious to eat, as delicious a tapa as it is a side dish.

SERVES **4**

- 1½ cups freshly cooked puy or other dark, small, firm lentils, or 1 cup uncooked puy lentils with 3 bay leaves (plus water to cover)
- 3–4 Tbsp extra-virgin olive oil, plus extra to drizzle, if desired
- 3 garlic cloves, chopped
- ½ cup tomato juice
- 3 ripe tomatoes, chopped (if using canned tomatoes, include their juices and omit the tomato juice)
- ¾ cup Spanish chorizo sausage, cut into small pieces
- 2–3 tsp chopped, fresh rosemary
- Salt and pepper to taste

Preparation: 15–20 minutes

Cooking time: 1 hour

❶ If the lentils are uncooked, place them in a saucepan with the bay leaves and water. Bring them to a boil, then reduce the heat and leave them to simmer until they are tender, about 30 minutes. If they are still a little firm, remove them from the heat and leave them for another 30 minutes in the hot water, covered, to plump up.

❷ Heat the olive oil and garlic in a large skillet, and when fragrant, add the lentils, plus three fluid ounces of their cooking liquid and the tomato juice. Cook over a high heat until the liquid is nearly all evaporated, then add the tomatoes, chorizo, and rosemary.

❸ Cook together for a few minutes, then taste for seasoning, and if needed, add salt and pepper to taste. Serve immediately, with an extra drizzle of olive oil, if desired.

Stewed Peas and Asparagus

piselli e asparagi

When young, tender rutabaga are available, I add three to four, blanched and cut into bite-size pieces, to the vegetable mix.

SERVES 4

- I onion, chopped
- 5 garlic cloves, chopped
- 4 Tbsp extra-virgin olive oil
- Salt and pepper to taste
- 1½–2 cups/8–12 oz peas
- 2 cups/14 oz ripe, diced tomatoes, or I can tomatoes including juice
- I tsp chopped, fresh rosemary
- 15–20 asparagus spears
- Large pinch of saffron, dissolved in I Tbsp of warm water

Preparation: 20 minutes

Cooking time: 10–15 minutes

❶ Lightly sauté the onion and half the garlic in the olive oil until they are tender, then season with salt and pepper. Add the peas, stir for a few minutes, then add the tomatoes and rosemary, and cook for about five minutes.

❷ Cut the asparagus into bite-sized pieces and add to the pan with the saffron, and continue to cook for another five minutes or so, then either serve hot or let cool to room temperature, and enjoy when tepid.

Sautéed Celeriac

I enjoyed the following dish not long ago in France—blanched celeriac "chips," in olive oil and sprinkled with parsley. Simple, but embarrassingly "more-ish."

SERVES 4

- I medium-sized celeriac
- Cold water
- Juice of ½ lemon
- ½–I tsp sea salt, or as desired
- Black pepper to taste
- I cup all-purpose flour, or as needed
- ½ cup pure olive oil
- ¼ cup extra-virgin olive oil, or as needed; a good Greek or Spanish oil
- I Tbsp chopped, fresh parsley
- I Tbsp chopped, fresh chives

Preparation: 10–15 minutes

Cooking time: 10–15 minutes

❶ Peel the celeriac carefully with a sharp paring knife, and cut into chips or julienne strips, then submerge them in cold water. Add the lemon juice and leave until you are ready to sauté them, up to an hour.

❷ Mix the salt and pepper into the flour, and place it in a big bowl or plastic bag for dipping.

Drain the celeriac, pat it dry, then toss it in the seasoned flour in the bowl or bag, and mix to coat well. Remove it from the flour.

❸ Heat the oil in a skillet until it is hot but not smoking, and in several batches (depending on the size of the pan), cook the floured celeriac until it is golden on both sides. Remove from the hot oil, drain briefly on paper towels, sprinkle with chopped fresh parsley and chives, and serve immediately.

▲ *Stewed Peas and Asparagus*

Summer Squash Casserole

tourlu

Vegetable casseroles abound in Greece during the lush, fertile summer. Eggplants, peppers, tomatoes, and zucchini … especially zucchini and other summer squashes. This casserole of tomato-y, cooked vegetables, topped with a feta-cheese–soufflé-like layer, can be made with any summer vegetables. Eggplant is also very nice. This dish is good warm or cold.

SERVES 4

- 2–3 Tbsp extra-virgin olive oil, preferably Greek
- 2 lbs zucchini or mixed summer squash, cut into bite-sized pieces
- 3–5 garlic cloves, coarsely chopped
- 2 cups/14 oz diced tomatoes
- 1–2 Tbsp tomato paste, if needed
- Salt, pepper, and a pinch of sugar, to taste
- ¼ teaspoon oregano, or to taste
- 4 eggs, lightly beaten
- 3 fl oz milk
- 2 cups feta cheese, cut into bite-sized pieces, plus the crumbs
- Grated nutmeg, to taste

Preparation: 15–20 minutes

Cooking time: 1 hour

❶ Sauté the onion in the olive oil until it softens, then add the zucchini or squash and half of the garlic, and stir until lightly cooked. Add the tomatoes and cook for about ten minutes or until a tomato sauce is formed. Stir in the tomato paste if needed, and season with salt, pepper, a pinch of sugar, and oregano to taste. Pour into a casserole or baking dish large enough for the mixture to fill to about half-full.

❷ Combine the eggs with the milk, feta cheese, nutmeg, and pepper to taste. Pour this over the zucchini or squash and tomatoes.

❸ Bake uncovered in a preheated (375°F) oven for about 30 minutes, or long enough for the casserole to puff slightly and turn golden brown on the top. Remove from the oven.

❹ Serve warm, or cool if preferred.

vegetable dishes

74

Rich and flavorful, olive oil is unsurpassed for cooking fish, meat and poultry, whether used to stew, braise, sauté, or roast. It gives more flavor and character than other oils, and doesn't have the heaviness of butter.

Roasted Chicken Breasts

served with roasted red peppers and tomatoes with broiled asparagus

A little marinade of olive oil does chicken breasts a world of good. It keeps the meat so moist and juicy, and gives an effortlessly Mediterranean flavor.

SERVES 4
- 4 chicken breasts, boned
- 3–4 Tbsp extra-virgin olive oil
- 1 tsp balsamic vinegar
- Salt, pepper, and Herbes de Provence, Bouquet Garni, or thyme, to taste
- 5 garlic cloves, chopped
- 1 bunch of thin asparagus
- 1 Tbsp white wine
- 2 red bell peppers, roasted and peeled
- 4 small, ripe tomatoes, cut into halves crosswise

Preparation: 20 minutes

Cooking time: 15–20 minutes

❶ Combine the chicken breasts with two tablespoons of the olive oil, the balsamic vinegar, salt, pepper, and Herbes de Provence, and a third of the garlic. Leave to marinate while you prepare the other ingredients.

❷ Combine the asparagus with a tablespoon of the olive oil, the wine, salt, pepper, Herbes de Provence, and a third of the garlic. Place on a baking sheet.

❸ Heat a heavy skillet, and when it is very hot, add the chicken breasts and sear on both sides,

then lower the heat, and add the peppers and tomatoes. Each should be seared, but not overcooked. As each is ready, remove from the pan.

❹ Meanwhile, broil the asparagus until they are lightly browned and still crispy-tender, but crunchy. Turn once or twice.

❺ Serve the chicken, peppers, tomatoes, and asparagus with the remaining olive oil and garlic sprinkled over it.

fish, meat, and poultry

Moroccan Chicken

tajine msir zeetoon

This is one of my favorite dishes, as the lemon permeates the chicken, and the olives add piquancy. There is little work in preparing it—it basically simmers away in the oven—and it is never less than delicious. Any type of olives are delicious here; or you can use only one type if you like, but I prefer to use three different types of olives for variety.

SERVES 4

- 1 chicken, cut into serving pieces
- 1 Tbsp cumin
- 2 tsp paprika
- ½–1 tsp ginger
- ½–1 tsp turmeric
- 5 garlic cloves, chopped
- Several handfuls of fresh cilantro, chopped
- Juice of 2 lemons
- Black and cayenne pepper, to taste
- 3–5 Tbsp flour
- 4 tomatoes, chopped (either ripe or canned)
- 10–15 each (three types in total): green olives of choice, black olives, cracked olives, oil-cured olives, purplish-red olives, Kalamata, pimiento-stuffed green olives, etc., drained
- 1 lemon, cut into 6 wedges
- ¼ cup extra-virgin olive oil
- 1 cup chicken broth
- Extra lemon juice, to taste

Preparation: 10 minutes

Marinating time: 30 minutes

❶ Combine the chicken with cumin, paprika, ginger, turmeric, garlic, cilantro, lemon juice, and pepper, and place in a baking dish in a single layer. Leave to marinate for 30 minutes, then add the flour, and toss together to coat well.

❷ Heat the oven to 325°F. Add the chopped tomatoes, olives, lemon wedges, olive oil, and broth to the dish. Bake uncovered for about an hour, or until the chicken is tender and a delicious sauce has formed.

▲ Pan-Broiled Shrimp Kabobs

Pan-Broiled Shrimp Kabobs

with roasted tomatoes

With such simple ingredients, the quality of the olive oil is very important: it should sing with flavor! Ditto for the tomatoes.

This may be prepared on the outdoor grill if you like, and it is actually a very convenient dish to prepare for entertaining, as the tomatoes not only may be roasted the day ahead, but they are at their best that way.

SERVES 4

- 8–12 small-to-medium, ripe, flavorful tomatoes
- Pinch of sugar
- Pinch of salt
- 45–50 small-to-medium shrimps and about 35–40 jumbo shrimps, in their shells, heads and tails can be removed (optional)
- ⅓–½ cup/3–4 fl oz extra-virgin olive oil
- 2 tsp balsamic vinegar
- 3 garlic cloves, chopped, or to taste
- Salt and a few grinds of black pepper, to taste
- 2–3 Tbsp fresh basil leaves, thinly sliced or torn
- Lemon wedges, for serving

Preparation: 30 minutes

Marinating time: 30 minutes

Cooking time: 1 hour

❶ Place the tomatoes in a roasting pan, preferably a ceramic Mediterranean one, then bake in the oven at 375°F, uncovered, for 20–30 minutes. The skin should have split, exposing some flesh. Sprinkle with sugar and salt, then return to the oven and continue to roast for another 15–25 minutes.

❷ Remove and let cool. It is best to let them sit overnight, as the juices will run out and thicken.

❸ Remove the skins of the tomatoes, and squeeze them to extract their flavorful juices. Discard the squeezed-out skins, and pour the juices over the roasted tomatoes, then cut the tomatoes into halves or quarters.

❹ Place the shrimps in a nonreactive bowl for marinating; add several tablespoons of olive oil, a teaspoon of balsamic vinegar, and half the garlic. Leave for at least 30 minutes. Meanwhile, soak 8–12 bamboo skewers in cold water (or use metal skewers to avoid soaking).

❺ Tightly thread the shrimp onto the skewers. Save the marinade to heat through as a pan sauce.

❻ Heat a skillet and brown the shrimp quickly on each side, for only a few minutes, depending on their size. Remove to a plate, and keep warm.

❼ Heat the tomatoes in the pan, then remove to the plate, and sprinkle with the remaining garlic. Pour the marinade into the pan, heat through until it bubbles, then pour over the shrimp skewers and tomatoes. Sprinkle with salt and pepper, then the basil and serve right away, accompanied by lemon wedges.

fish, meat, and poultry

Caribbean Seared Tuna

A wonderful example of a small amount of oil, which gives a lovely, large amount of flavor. Serve with sweet potatoes brushed with olive oil, then broiled or roasted.

SERVES 4

- 4 tuna steaks, about 6 oz each
- 3 garlic cloves, chopped
- ½ tsp salt
- ½ tsp cumin
- Juice of 2 limes, or juice of 1 orange and 1 lime
- Black and cayenne pepper to taste
- 2 Tbsp extra-virgin olive oil, or as needed
- Cuban Garlic-citrus Sauce (pg. 121)

Preparation: 10 minutes

Marinating time: 30 minutes

Cooking time: 5–10 minutes

❶ Combine the tuna with the garlic, salt, cumin, lime juice, black and cayenne pepper. Leave for about 30 minutes, then remove from the marinade and pat it dry.

❷ Brush the fish generously with olive oil, then cook quickly either in a skillet or over a charcoal fire, about a minute or two on each side only.

❸ Serve with the sauce spooned over it when hot.

Mediterranean Poached Fish

The leftover juices from this dish make a superb broth—serve Greek style, thickened with beaten, raw egg (see note, page 4) and lemon.

SERVES 4

- 1 large onion, chopped
- 2 Tbsp fresh, chopped parsley
- 1 bay leaf
- Several sprigs of thyme
- Pinch of cloves
- Salt and pepper to taste
- 1 cup dry white wine
- 1 cup fish broth
- 1 cup water
- 1 lb small, new potatoes, peeled
- 2½ lbs fish of choice for poaching
- ½ cup extra-virgin olive oil
- Juice of 1 lemon

Preparation: 30 minutes

Cooking time: 1hour

❶ Combine the onion, half the parsley, the bay leaf, thyme, cloves, salt, pepper, wine, fish broth, and water, and bring it to a boil. Reduce the heat, simmer for about 30 minutes, and strain.

❷ Place the potatoes in a saucepan, and pour half the strained broth over them. Bring to a boil, then reduce the heat, and let simmer until they are tender, about 15 minutes (depending on the size of the potatoes).

❸ Meanwhile, place the fish in the pan, and pour half of the hot broth over it. Then place on a low-to-medium heat, and let the fish and broth come to a gentle simmer (but do not boil). Just before the liquid boils, remove it from the heat, and let it sit for about 10 minutes. It should steep, rather than cook forcibly.

❹ Add the hot, boiled, drained potatoes to the hot, drained fish (reserving liquids from both to make a soup), drizzling with the olive oil, and squeezing the lemon juice over it. Sprinkle with the reserved parsley, and serve.

▲ Caribbean Seared Tuna

Monkfish with Orange,

tarragon, garlic, and bay leaves, with a flame of brandy

I adore this flavoring combination for a wide variety of foods—pork, chicken, and shellfish, especially. Try scallops or shrimps in place of the monkfish, or make a melange of shrimps, scallops, and monkfish.

SERVES 4

- 1½ lbs monkfish steaks
- 8 oz shrimp, left in their shells for maximum flavor, or removed from their shells for easier eating
- 1–2 tsp fresh tarragon leaves
- 4–6 bay leaves, broken in halves
- 10 garlic cloves, coarsely chopped
- 1 cup dry white wine
- 1 cup orange juice
- Grated rind of ½ orange
- Salt and pepper to taste
- ¼ cup/2–3 fl oz extra-virgin olive oil
- 2 Tbsp brandy
- 1 Tbsp fresh, chopped parsley

Preparation: 15–20 minutes

Marinating time: 1–3 hours

Cooking time: 10–15 minutes

❶ Combine the monkfish and shrimp with the tarragon, bay leaves, garlic, wine, orange juice, orange rind, salt, and pepper. Leave this to marinate for at least an hour, or up to three hours in the refrigerator.

❷ Remove the fish and shrimp from their marinade, putting the marinade aside and pat dry.

❸ Heat a heavy skillet, preferably nonstick, with the olive oil, and when it is very hot, add the shrimp and monkfish, only enough so that they do not crowd the pan. Cook on one side until they just begin to change color, then turn to the other side, and lightly brown them. You want them to stay juicy and not overcook—allow only a few minutes in total. Pour the brandy in the pan; it will flame up quickly, so be extremely careful and take care of your face, eyebrows, and any curtains near the stove. When the flames die down, remove the shrimp from the pan.

❹ Add the marinade and reduce down until it forms a flavorful sauce, about five to eight minutes, then return the monkfish and its juices to the pan (discarding the bay leaves). Warm through, sprinkle with the chopped parsley, and serve.

fish, meat, and poultry

Picadillo

Picadillo is a Latin American mixture of ground meat, browned and simmered with spicy–sweet–savory ingredients. It is delicious rolled into flour tortillas, or covered with a layer of puff pastry for a large pie, or—the more labor-intensive version—as a filling for individual little pastries or roasted, large, green chiles.

The hallmark flavors of picadillo are sweet raisins, crunchy nuts, and saline, pungent olives, with a good shot of cinnamon and sugar added to the mix.

SERVES **4**

- 1 onion, chopped
- 3 garlic cloves, chopped
- 2 Tbsp extra-virgin olive oil
- 1 lb lean ground beef
- ⅛–¼ tsp cinnamon
- ⅛–¼ tsp cumin
- ⅛–¼ tsp cayenne pepper
- Pinch of cloves
- 4 heaped Tbsp raisins
- 4 Tbsp toasted almonds or cashews
- 3 fl oz sherry or dry, red wine
- 2–3 ripe tomtoes, diced
- 10–15 pimiento-stuffed green olives, sliced or halved
- 2 Tbsp tomato paste
- 1–2 Tbsp sugar, to taste (this is a sweet-and-sour sauce)
- 1–2 Tbsp red-wine or sherry vinegar, to taste
- 2 Tbsp chopped, fresh cilantro

Preparation: 20 minutes

Cooking time: 30 minutes

❶ Sauté the onion and garlic in the olive oil until soft, then add the beef and brown, sprinkling with the cinnamon, cumin, cayenne pepper, and cloves as it cooks.

❷ Add the raisins, almonds or cashews, sherry or red wine, and bring to a boil. Cook until the sherry/wine has nearly

evaporated, then add the tomatoes, olives, tomato paste, sugar, and vinegar, and cook together until the sauce is thick and flavorful.

❸ Stir in the cilantro, and serve as desired.

Leg of Lamb – French-style

gigot à l'ail aux olives

Greed wells up in me with a mad passion when I think of roasted lamb with whole garlic cloves and olives. Sometimes, just for the sheer delight of it, I toss a few artichokes into the roast as well. Aioli (pg. 120) is a nice accompaniment to this dish.

SERVES 6

- 4 lb leg of lamb
- Rosemary or thyme, fresh or dried, as desired
- 8 garlic cloves, cut into slivers
- 2 heads of garlic, with cloves separated but not peeled
- 4–6 Tbsp extra-virgin olive oil
- Salt and pepper as desired
- 1 cup chicken broth
- 1 cup red wine
- Several handfuls of flavorful olives of choice, either green or black
- 1–2 Tbsp fresh, chopped parsley or other fresh herbs, such as marjoram

Preparation: 15–20 minutes

Cooking time: 3 hours (approx.)

❶ Make incisions all over the lamb, and into each incision place a pinch of thyme or rosemary and a sliver or two of garlic. Place the lamb in a roasting pan and surround it with the whole garlics. Drizzle four tablespoons of olive oil over the lamb, and sprinkle with any leftover, slivered garlic, salt, and pepper.

❷ Roast, uncovered, at 350°F in the oven until the lamb reaches an inner temperature of 125°F. The lamb should be pink. This is an excellent time to use a meat thermometer, to check the meat's internal temperature.

❸ Remove from the oven and place the meat on a platter to keep warm. Skim the fat from the pan, and add the broth and red wine. Boil down until the sauce has reduced by about half and is very flavorful, then add the olives, and warm through.

❹ Serve the lamb sliced and surrounded by the roasted garlic-and-olive sauce, and a sprinkling of chopped parsley or herbs.

▲ *Leg of Lamb—French-style*

Beef with Provençal Flavors

daube provençal

This may be made with any good, stewing cut: cheeks of beef are very fashionable these days, but chuck is more easily available.

Marinating the meat infuses it with flavor, while the wine helps to tenderize it. This is also a marvelous way to stew lamb. Black olives are delicious in this, but so are green. Whichever olives you choose, each will give their own distinctive character to the dish.

SERVES 6

- 1 carrot, chopped
- 2 leeks, chopped
- 1 bottle of red wine, such as Côtes du Rhone or Côtes de Provence
- 1 head garlic, cut in half crosswise
- 2½ lbs boneless, stewing beef, such as chuck, cut into chunks
- 2 bay leaves
- 1 tsp fresh (or several pinches of dried) thyme
- 4–5 Tbsp extra-virgin olive oil
- 1 red bell pepper, diced
- 1–2 tsp Herbes de Provence or Bouquet Garni
- Grated rind of ¼ orange, chopped or in one piece
- 1 cup beef broth
- 2–3 Tbsp cognac or brandy
- 1⅛ cups diced tomatoes (canned tomatoes are fine)
- 3–5 garlic cloves, chopped
- 1–2 Tbsp chopped, fresh parsley
- 10–15 black and 10–15 green olives; use good, Mediterranean, flavorful olives of choice

Preparation: 30 minutes

Marinating time: at least 5 hours

Cooking time: 3–4 hours

❶ Combine the carrot, leek, wine, garlic halves, and beef in a bowl, and add the bay leaves and thyme. Leave to marinate for at least five hours or preferably one or two nights in the refrigerator.

❷ Remove the meat from its marinade, and let it sit a few minutes, preferably in a strainer, to remove its excess liquid. Pat the meat dry with paper towels.

❸ Remove the chopped vegetables from the marinade and strain in a colander, discarding the muddy sediment at the bottom of the bowl.

❹ Heat a few tablespoons of the olive oil in a heavy casserole, and lightly sauté the carrot, leeks, and garlic. Remove for a moment, then brown the meat and the red pepper, in small batches, adding more olive oil as needed.

❺ Layer the sautéed vegetables and meat in the casserole, and add the strained marinade, the Herbes de Provence, grated orange rind, broth, cognac or brandy, and the tomatoes. Bring to a boil, then cover and reduce the heat to very, very low, or place it in a preheated (325°F) oven, and leave to slowly cook until the meat is very tender, about three to four hours.

❻ Carefully pour off the sauce, skim off the fat, then pour it into a saucepan, and reduce until intensely flavorful. This may take about 15 minutes.

❼ Taste for seasoning, then stir the sauce back into the meat stew, adding the garlic, parsley, and olives, and warm through. Serve immediately. Like most stews, though, this is even better the next day.

Pasta and olive oil—these two ingredients are practically a recipe in themselves. Their basic goodness acts with whatever else you choose to give you a delicious, succulent, and healthy meal. Rice and bulgur wheat, too, are delicious and savory, when cooked with olive oil.

pasta and grains

Pasta with Raw Tomato Sauce,

basil, and goat's cheese

This recipe is the classic combination of hot pasta with raw tomatoes, rich with olive oil, basil, garlic, and a crumbling of goat's cheese for fresh piquancy found in the streets of Naples—a true Italian tradition.

SERVES 4

- 10 very ripe, juicy tomatoes, diced
- Salt, to taste
- Few drops of balsamic vinegar
- 3 garlic cloves, chopped
- 4–6 Tbsp extra-virgin olive oil
- Several handfuls of fresh basil, torn coarsely
- Few pinches of chili flakes (optional)
- 12 oz pasta of choice
- 1 cup goat's cheese, crumbled

Preparation: 10 minutes

Marinating time: 1½–2 hours

Cooking time: 10 minutes

❶ Combine the diced tomatoes with several pinches of salt, balsamic vinegar, chopped garlic, most of the olive oil (except one tablespoon for tossing the cooked pasta) and basil (plus chili flakes, if using). Leave to sit for at least 30 minutes, or preferably chill in the refrigerator for several hours.

❷ When ready to serve, cook the pasta until *al dente* (firm to the bite), then drain and toss with the goat's cheese and olive oil. Add to the sauce, and serve immediately.

Pasta with Tuna and Tomato

pasta con il tonno rosso

The addition of saline, juicy chunks of tuna is a classic of the Italian kitchen. Don't worry about using canned tuna as a substitute for fresh, as the canned is actually more authentic, though you could use leftover chunks of grilled tuna in its place.

SERVES 4–6

- 1 onion, chopped
- 5 garlic cloves, chopped
- 3 fl oz extra-virgin olive oil, or as desired
- 2½ lbs diced tomatoes (or one 14-oz can, including the juices)
- 2 Tbsp tomato paste (concentrated purée), if needed
- Salt, pepper, and a pinch of sugar if needed to balance the acidity
- Pinch of oregano and/or marjoram, to taste
- 1 can tuna, 6½ oz (approximately), drained
- 1 lb spaghetti
- 3 Tbsp fresh, chopped parsley
- 3 Tbsp capers, preferably salted rather than brined, but rinse well if brined
- 10–15 black olives, such as Gaeta, pitted and quartered
- Few Tbsp of toasted bread crumbs (optional)

Preparation: 15–20 minutes

Cooking time: 10–15 minutes

❶ Lightly sauté the chopped onion and half the garlic in the olive oil until softened, then add the tomatoes and cook over medium heat until the tomatoes are juicy. Add the tomato paste if needed, then season with salt, pepper, sugar, and a pinch of oregano or marjoram to taste. Add the tuna to the sauce, and heat through.

❷ Meanwhile, cook the pasta in rapidly boiling, salted water until *al dente* (firm to the bite), then drain. Pour half the sauce into the pasta, and toss together with the capers and olives, then toss with the rest of the sauce. Sprinkle with the fresh, chopped parsley and the toasted bread crumbs (if using), and serve immediately.

pasta and grains

95

Spaghetti with Tomato and Clams

spaghetti alla vongole

The first time I ate this, I was on my first foray to the continent, and on discovering the old cobbled streets of Naples, promptly sat down and ordered dinner. When it arrived, I had never seen anything so beautifully delicious—tiny clams tossed on a sea of tomato-based spaghetti sauce, with a dusting of fine, green parsley and the heady scent of garlic.

Choose tiny clams, if you like—they look lovely, but you must be careful not to swallow the shells. Larger clams are more convenient; if clams are not available, use mussels.

SERVES 4

- 2¼ lbs fresh clams in their shells, or mussels (removed from shells)
- 8 garlic cloves, chopped
- ½ cup fruity, strong-flavored, extra-virgin olive oil
- Several pinches of hot, red-pepper flakes, or ½–1 dried, red chile, crushed
- ½ cup dry white wine
- 1 cup puréed tomatoes
- 1 Tbsp tomato paste (double purée)
- Sea salt to taste, and a pinch of sugar to balance the acidity, if needed
- Several large pinches of dried oregano, crumbled
- 1 lb spaghetti
- 2–3 Tbsp fresh, chopped parsley

Preparation: I hour, 10 minutes

Cooking time: 15–20 minutes

❶ Cover the clams or mussels with cold, salted water, and leave them for 30–60 minutes to clean. Remove and drain.

❷ Sauté the clams or mussels with half the garlic in the olive oil for about five minutes, then add the hot-pepper flakes and wine, and cook over high heat to evaporate.

Add the puréed tomatoes, tomato paste, remaining garlic, salt, sugar, and oregano. Cover and cook over medium heat until the clams pop open, about 10 minutes.

❸ Cook the spaghetti until *al dente* (firm to the bite), then drain and toss with a few tablespoons of the sauce, pour it onto a large platter or into bowls, and top with the remaining sauce and clams. Sprinkle with fresh chopped parsley and serve.

pasta and grains

▲ *Spaghetti with Tomato and Clams*

Chicken and Olive Risotto

sbarrigia

Full of chicken, tomatoes, and olives, this rice dish is a classic of the Sicilian table.

SERVES 4

- ½ onion, chopped
- 4 garlic cloves, chopped
- 3 Tbsp extra-virgin olive oil
- 2 cups arborio rice
- 12–14 oz diced tomatoes, fresh or canned, with their juices
- ½ cup dry, white wine
- 1¾ pints chicken broth, or as needed (a bouillon cube or two mixed with water is fine)
- 25 green olives, pimiento-stuffed, or about 15 black and 15 green olives, pitted and cut into halves
- 1 chicken breast, poached, and cut into shreds or small pieces
- Several pinches of mixed Italian herbs, or a combination of rosemary, thyme, and sage
- Freshly ground black pepper for sprinkling (the olives and the Romano are both salty, so you probably won't need salt)
- 1 cup/3–4 oz Romano or other sharp cheese, grated coarsely
- 1–2 Tbsp chopped, fresh parsley

Preparation: 15–20 minutes

Cooking time: 20–30 minutes

Lightly sauté the onion and garlic in the olive oil until softened, then stir in the rice and cook for a few minutes until lightly golden. Add the tomatoes and cook for a few more minutes, then pour in the wine, stirring until the rice has absorbed the wine. Then, begin to add the broth, a few half-cups at a time, letting the rice absorb while you stir, and increasing the amounts while the rice cooks to absorb the liquid more quickly. When the rice is *al dente* (firm to the bite), stir in the olives, chicken, herbs, pepper, cheese, and parsley. Serve immediately.

Bulgur-wheat Pilaf

pourgouri pilafi

In our favorite Cypriot village taverna, at some point in the lavish meze dinner, a plate of this pilaf would make an appearance. You really do need good, fresh tomatoes for this dish.

SERVES 4

- 1 medium-large or 2 small onions, chopped
- 2 garlic cloves, chopped
- 3 medium-sized, ripe tomatoes, peeled and diced
- 2 Tbsp extra-virgin olive oil
- Handful of vermicelli, broken into small pieces (about 1 oz)
- 2 cups bulgur or cracked wheat
- 1½ cups vegetable or chicken broth, as desired
- Salt, black pepper, and crushed, dried oregano leaves, to taste

Preparation: 15 minutes

Cooking time: 10–15 minutes

▲ *Bulgur-wheat Pilaf*

❶ Lightly sauté the onion, garlic, and tomatoes in the olive oil until they have softened, then stir in the vermicelli pieces. It should be of a sautéeing consistency rather than a sauce consistency, so add more olive oil if needed.

❷ Add the bulgur or cracked wheat, stir, then add the broth, salt, pepper, and oregano. Cover and cook over a low heat until the broth is absorbed, and the wheat is chewy and tender, about 8 to 10 minutes.

❸ Leave the pilaf to sit a few minutes, then fluff it up with a fork and serve immediately.

pasta and grains

Arabic Rice and Lentils

immjadara

I prefer to use slate-gray-colored puy lentils for this dish, as they stay together so nicely, and have such a fine, earthy flavor.

Cooking in olive oil enhances the desert flavor of this simple, nourishing dish. Serve on its own or as a bed for grilled lamb kabobs and/or slices of grilled eggplant (pg. 59), and accompany with a bowl of cooling, unsweetened yogurt, a plate of sliced cucumbers, and a spicy relish, such as a Turkish-style red pepper sauce.

SERVES 4

- ¾ cup brown lentils
- Bay leaf
- 2 onions, thinly sliced
- 3 Tbsp extra-virgin olive oil
- 2 cups rice
- 5 cloves garlic, thinly sliced
- ½ tsp or more cumin
- 2 cups broth, either vegetable, chicken, or beef
- Several generous pinches of curry powder
- Seeds from 3–5 cardamom pods
- 2 garlic cloves, chopped fine
- Cayenne and black pepper, to taste
- Salt, if needed

Preparation: 15 minutes

Cooking time: 30–40 minutes

❶ Place the lentils in a saucepan, with the bay leaf and cover with water (water should cover the lentils by 1 inch). Bring to a boil, reduce the heat, and simmer, covered, for about 20 minutes or until the lentils are tender. Add more water if they seem too dry, or threaten to burn.

❷ Sauté the onions in the olive oil until they are lightly browned, then add the rice and cook with the olive oil and onions until lightly golden. Stir in the garlic and cumin, cook for a few minutes longer, then stir in the broth, curry powder, and cardamom.

❸ Cover and cook over a medium heat until the rice is tender, for about eight minutes, depending on which rice you are using. After about six minutes, check the rice for liquid: if it seems too wet, remove the lid; if it seems too dry, add a little more water.

❹ When rice is tender, stir in the chopped garlic, cayenne and black peppers, and salt, if needed.

pasta and grains

I've included egg dishes in this chapter, dishes you might have for brunch or supper or indeed for a midnight feast. Fewer partners are more agreeable than eggs and olive oil—indeed, after cooking an omelet in olive oil, you may prefer it to butter forever.

Tunisian Scrambled Eggs

ajja

Spicy sausages flavor this dish of scrambled eggs with onion, peppers, and tomatoes, all cooked nutritiously in olive oil. For a vegetarian variation, I've made this dish using diced, smoked tofu instead of sausages, adding a mild chili powder, such as ancho or pasilla, for spice.

light supper and brunch dishes

SERVES 4

- 6–8 oz spicy sausages, such as merguez or Spanish chorizo (Mexican chorizo is too high in fat for this dish)
- I onion, chopped or thinly sliced
- I green bell pepper, thinly sliced
- I red bell pepper, thinly sliced
- 4–5 Tbsp extra-virgin olive oil
- 5 garlic cloves, chopped
- 5 ripe tomatoes, chopped or diced, including their juices (or use about 8–10 oz canned tomatoes)
- 6 eggs, lightly beaten (see note page 4)
- ¼ tsp cumin
- Salt and black pepper, to taste
- ¼–I fresh, green chile pepper, chopped, or to taste (or use a chile-garlic paste, such as a Turkish, Chinese, or Moroccan one)
- 2 Tbsp fresh, chopped cilantro

Preparation: 15–20 minutes

Cooking time: 15–20 minutes

❶ Slice the sausages and cook them in an ungreased pan until lightly browned, then pour off any excess fat, and drain on a plate or separate pan.

❷ Sauté the onion, and the red and green bell peppers in the olive oil until softened, then stir in the garlic and cook through for a few moments. Add the tomatoes, increase the heat, and cook over a medium high heat until the sauce is thick, about eight minutes.

❸ Pour in the eggs, sprinkle with cumin, salt, pepper, and chile, or chili paste as desired. Cook over a low-to-medium heat, stirring occasionally with a wooden spoon so that soft curds form as the beaten eggs cook.

❹ When the eggs are no longer runny, sprinkle with cilantro and serve immediately.

Rolled Omelet

with green chile, fresh dill, cilantro, and goat's cheese

The goat's-cheese filling is tangy and fresh, and the eggs are delicate when gently cooked in the fragrant olive oil.

SERVES **4**

- 8 eggs, lightly beaten
- 2–3 Tbsp milk
- Salt and pepper, as needed
- 4–5 Tbsp extra-virgin olive oil
- ½ or more green chile, chopped fine
- 2 garlic cloves, chopped fine
- 4–6 oz goat's cheese
- 3 Tbsp chopped dill
- 3 Tbsp fresh, chopped cilantro
- Sour cream to taste (optional)

Preparation: 10–15 minutes

Cooking time: 10–15 minutes

❶ Prepare four individual omelets, using two eggs for each, mixed with a little milk per omelet. Pour a tablespoon or two of olive oil into the omelet pan, and warm but don't overheat—it should smell fragrant, and be almost smoking, but not quite.

❷ Pour in the required amount of beaten egg, and cook for a few moments over a low heat, lifting the edges from the sides, and letting the runny egg flow under. When the egg is nearly set, sprinkle in a quarter of the chile, garlic, goat's cheese, dill, and cilantro, and fold over.

❸ Roll out of the pan and serve hot, with each omelet topped with a scoop of sour cream, if desired.

Green-Bean Frittatas

with olivada topping

Tender, young, thin, green beans, cooked in a flat omelet in olive oil, then topped with a thin layer of black-olive paste (olivada) or tapenade, makes a delicious dish for picnics, or served as an antipasto or tapa.

SERVES 4

- 1 cup thin or extra-fine green beans
- 3 garlic cloves, minced
- 3–4 Tbsp extra-virgin olive oil, preferably Spanish, Provençal or Greek, or as needed
- Salt and pepper
- 6 eggs, lightly beaten
- 3 Tbsp olive paste (olivada or tapenade, as much as desired)
- 2 Tbsp thinly sliced, fresh, sweet basil

Preparation: 15–20 minutes

Cooking time: 15 minutes

❶ Steam or blanch the green beans until they are bright green and crispy-tender. Rinse them in cold water to stop their cooking. Strain, and then cut them into bite-sized lengths.

❷ Warm the garlic in about a tablespoon of the olive oil, then add the green beans and warm through with the garlic. Season with salt and pepper. Remove the green beans from the pan and add them to the eggs, stirring to mix well with the eggs.

❸ In the same pan, heat a tablespoon or so of the olive oil, and when it is hot (but not smoking), add a quarter of the green bean–egg mixture, pouring and spreading it so that it forms a flat omelet. Cook over a medium heat until the bottom is golden, then brown the top in the broiler. Remove it and place on a baking sheet. Repeat until you have four little, flat omelets.

❹ Spread each omelet with olive paste, then sprinkle with basil.

light supper and brunch dishes

Olive oil with bread forms the basic meal for much of the Mediterranean. Salad, a little broiled meat, an omelet, or a plate of pasta, are all mere additions to the basic everyday food of bread and oil.

For this most Mediterranean of foods, drizzle the olive oil on a good, hearty bread. Sprinkle it with a coarse sea salt, and if the season is right, add a layer of fresh, sweet, ripe tomatoes and let their juices sink in. This is often my breakfast, with an added sprinkling of fresh garlic and basil, just a little extra olive oil, and perhaps an olive or two on the side.

Cuban Sandwich

A Cuban sandwich is a soft French-like loaf or baguette, filled with various meats, cheeses, and so forth, then cooked in a skillet, often with a heavy weight pressed on it. The result is a crisply-crusted, savory-filled sandwich, at its best when opened and stuffed with a few little leaves of crunchy, fresh lettuce, and a slice or two of ripe tomato.

SERVES 4
- 1 soft stick French bread (half-baked is best)
- Extra-virgin olive oil as needed (Spanish is preferable)
- Cuban Garlic-citrus Sauce (pg. 121)
- 1 cup thinly sliced ham
- 1 cup thinly sliced, white cheese, such as Monterey Jack, manchego, or cheddar
- 1 cooked chicken breast (4–6 oz) or cooked turkey breast, sliced thinly (alternatively, any cooked meat, such as pork, beef, or lamb,) thinly sliced
- 10–15 pimiento-stuffed green olives, sliced or halved
- Few leaves of green lettuce
- 2–3 ripe, sliced tomatoes

Preparation: 10 minutes

Cooking time: 15 minutes

❶ Brush the bread outside and inside with the olive oil, then splash it with a little of the mojo sauce. Layer the ham, cheese, chicken, or meat, and olives inside, splashing them all with a bit of mojo, then close up and press together tightly. Cut the loaf into four individual sandwiches.

❷ Heat a skillet and brown the sandwiches in the pan; if you have a heavy weight, place this on top of the sandwich to keep it flat as it browns. Brown on both sides until the sandwich is crusty and brown, and the cheese has melted, then open the sandwich and tuck in the lettuce and tomato. Serve immediately, with more sauce if desired.

VARIATION

Place the sandwiches on a baking sheet and bake at 400°F until browned and the cheese has melted. Press the sandwiches down when you turn them over, and serve when sizzling hot.

pizzas, sandwiches, and savories

Goat's Cheese Tartina

tartina al caprina

Nothing could be simpler than fresh, tangy goat's cheese, spread onto bread and dribbled with olive oil infused with fresh herbs and garlic, but few things taste better.

SERVES 4
- 3 garlic cloves
- Several pinches of salt
- 1 tsp lemon juice
- 1–3 tsp each tarragon, basil, parsley, dill, rosemary, and mint, or as desired
- 2 Tbsp fresh, chopped chives
- 3 fl oz flavorful extra-virgin olive oil, such as Provençal

- 8 small, thin slices of rye bread, preferably French- or German-style, without seeds, or French *pain levain*
- Fresh, mild goat's cheese, as desired

Preparation: 10–15 minutes

Marinating time: at least 1 hour

❶ Crush the garlic with a mortar and pestle, and add the salt, then work in the lemon juice. Add the herbs and chives, then stir in the olive oil and leave to marinate for an hour or longer.

❷ Spread the bread with the goat's cheese, then drizzle a little of the oil, either with the herbs intact or strained. Serve immediately.

109

Muffaletta

New Orleans-style sandwich

pizzas, sandwiches, and savories

SERVES 4–6

- 5 garlic cloves, chopped
- 25–35 black olives, pitted and sliced
- 25–35 green pimiento-stuffed olives, sliced
- 2 roasted, red bell peppers, peeled and cut into strips
- 3–5 Tbsp fresh, chopped parsley
- ½–¾ cup extra-virgin olive oil
- 1–2 Tbsp white-wine vinegar
- Several pinches of oregano
- 1 baguette or other freshly baked, country bread loaf
- 1–1½ cups/5–6 oz salami, sliced
- 1–1½ cups/5–6 oz mortadella, or more, sliced
- 1–1½ cups/5–6 oz prosciutto or Parma ham, or more, sliced
- 1–1½ cups/5–6 oz Spanish chorizo sausage, sliced
- 3 cups thinly sliced, mild cheese, such as Monterey Jack, fontina, etc.

Preparation: 20 minutes

❶ Combine the garlic with the olives, peppers, parsley, olive oil, vinegar, and oregano.

❷ Slice open the bread, and hollow out some of the center of the loaf.

❸ Drizzle the cut sides with some of the oily juices from the olive mixture, then fill the bottom with about two-thirds of the olive salad, then layer with the meats and cheeses, then top with the rest of the olive mixture. Close up tightly, and chill until ready to eat.

This New Orleans-style sandwich is nearly as legendary as the oyster po'boy from the same town. Grocery stores all over New Orleans sell muffalettas, and every resident has his or her favorite. A muffaletta is a big chunk of Mediterranean-style country bread, best slightly sour, if your tastes are anything like mine.

A muffaletta is as excessive as the city it comes from—filled with roasted peppers, a relish of green and black olives, lots of garlic, and a layering of salami, hams, mortadella, etc., with a few extra slices of cheese thrown in just to be sure.

Bocadillo from Santa Gertrudis

Santa Gertrudis is a little village in the hills of the Mediterranean island of Ibiza, which consists of an art gallery, a craft store, an automatic bank machine, and a bar that serves the most delicious little sandwiches, or bocadillos. For breakfast, we always like the one filled with ham and manchego cheese, but some have only tomato and garlic, or grilled steak around lunch time. With a good, chilled beer, it is perfection.

SERVES 4

- 4 large crusty bread rolls, with soft, tender, slightly-sour crumbs
- 4–8 garlic cloves, cut into halves
- Extra-virgin olive oil, to drizzle generously
- 2–3 ripe tomatoes, slightly squished or thinly sliced
- Sprinkling of oregano
- 8–12 slices Parma ham, or prosciutto
- 8–12 thin slices of manchego or other mild, white cheese (I use Monterey Jack, or Sonoma)

Preparation: 15 minutes

Cooking time: 10 minutes

❶ Partially cut, partially break apart the bread rolls lengthwise, and lightly toast. Rub the cut garlic onto the toasted cut side of the bread—the coarse edges are slightly sharp and will act as a shredder. If there is any garlic left, just chop and sprinkle it on.

❷ Drizzle the cut garlic-rubbed rolls with olive oil, then rub with the tomatoes and sprinkle with oregano. Layer with the ham and cheese, close up, and heat again, preferably in a pan, and with a heavy weight on the sandwich that will press it together as it heats through.

❸ Serve immediately, with a few olives on the side, and the aforementioned beer.

Green Olive

and red-pepper piadina

Piadina, occasionally available in some Italian delis, are flat, wheat-flour pancakes from Northern Italy near Bologna, where they are served with salami, cured meats, and cheeses. I find them nearly identical to flour tortillas, and often use the easily available Mexican pancakes for a piadina-type snack. This makes a marvelous starter; if you like, add a spoonful of salsa for a bit of heat and pizazz … but the soul and flavor of the dish lies in the olives, cheese, and drizzled olive oil.

SERVES 4

- 4 flour tortillas
- 12 pimiento-stuffed green olives, sliced
- 2 garlic cloves, minced
- 1 pickled jalapeno chile pepper (in jar) or ½ fresh, hot chile pepper, minced
- 1 red bell pepper, roasted, peeled, and cut into strips
- 8 oz thinly sliced, mild, white cheese, such as Monterey Jack, cheddar, or fontina
- 1–2 tsp extra-virgin olive oil
- 1 Tbsp freshly chopped cilantro

Preparation: 5–10 minutes

Cooking time: 5–10 minutes

❶ Arrange the tortillas on a baking sheet, and scatter the top with the olives, garlic, chile, and pepper. Top with the cheese, and drizzle with the olive oil.

❷ Grill until the cheese melts, and serve right away, with a sprinkling of cilantro.

Olive-batter Bread

eliopita

I have tinkered with the traditional eliopita, since I find that the pine-like scent of rosemary is a delicious improvement. While black olives are traditional, I make my eliopita with green, black, or a combination of olives, often making this Cypriot snack with the oil from oil-marinated olives—if it's not too salty, that is.

SERVES 4

- 3 cups all-purpose flour
- ½ cup whole-wheat flour
- 1 tsp baking soda
- 1 Tbsp dried-mint leaves
- 1 Tbsp fresh rosemary, chopped
- ½ cup extra-virgin olive oil
- 1 cup warm water
- 1 onion, chopped
- Approximately 20 olives, pitted and cut into quarters or small pieces

Preparation: 20 minutes

Cooking time: 40 minutes

❶ Combine the two flours with the baking soda and the mint, and set aside.

❷ Combine the rosemary, olive oil, and warm water, and stir into the flour mixture, then stir in the onion and olives. Work only until the mixture has the consistency of a thick batter.

❸ Pour the batter into a lightly oiled, 9 x 13-inch or 12 x 15-inch baking pan, and bake at 350°F for 30–40 minutes, or until evenly browned and firm. Though it is traditionally served hot, personally I like to eat it cool, after it has set a while. When hot, it has a tendency to be a bit squishy inside.

▲ *Green Olive and Red-Pepper Piadina*

Pizza Dough

One of the joys of making your own pizza is choosing your toppings. The other is its convenience: a batch of dough in your refrigerator can keep for about a week— simply punch it down every so often, than take it out when you are ready to bake.

MAKES 6 INDIVIDUAL PIZZA OR
2 LARGE PIZZA

- 1¼ cups warm water (105–115°F— it should feel slightly hotter than your wrist)
- 1 package active, dried yeast
- 1 tsp sugar
- ¼ cup extra-virgin olive oil
- 1 lb strong, white flour
- 1 cup whole-wheat flour
- 1 tsp salt

Preparation: 2 hours
Cooking time: 10–30 minutes
(depends on size of pizza)

❶ Combine the water with the yeast and sugar. When it starts to foam, after 5–10 minutes, add the extra-virgin olive oil.

❷ Combine the two flours, and set aside about a cupful of this mixture. Mix the rest of it with the salt in a big bowl, and make a well. Into this well pour the liquid, and using a fork at first, work into the dough. When it forms a sticky mixture, replace the fork with a big, wooden spoon, then use your hands.

❸ Using the leftover flours, dust a big board and turn the dough out on it. Dust your hands with flour and knead, continuing until your dough is smooth and elastic, with an almost-satiny sheen. You will need to keep flouring the board and your hands.

❹ The dough is ready when a finger pressed against the dough will spring back from the indentation, after about 10 minutes' worth of kneading, place it in a large, oiled bowl to rise.

❺ Cover the bowl with a damp cloth or plastic wrap, and leave to rise. For it to rise quicker, leave in a warm place such as a warm kitchen or cupboard (about 90 minutes will do), or in the refrigerator, where it will take about a day and a half (good to know for make-ahead dough).

❻ Punch the risen dough down, and let it rise again; this time, it will rise much more quickly, and will have become quite malleable and easy to work with.

❼ After the dough has risen once again, punch it down and press it out, pizza-fashion, stretching the dough as you go, and arrange it on a baking pan or pizza pan. It is now ready to be topped with the ingredients of your choice and baked.

❽ Preheat the oven to 400–450°F. Fill a ceramic baking pan with boiling water, then place it in the bottom of the oven. This helps to create a crisp crust in the same way that a pizza stone or bricks do. If you have a pizza stone or bricks, use those instead. Drizzle the topped pizza with extra olive oil, and bake in a hot oven for 15–25 minutes, depending on the size of the pizza(s) and ingredients in the topping.

BASIC TOPPING

A good, basic topping is puréed tomatoes, with a scattering of diced, fresh tomatoes added for texture. Chopped garlic, oregano, and a layer of cheese if desired, and always a drizzle of a good, strong flavorful olive oil.

Pizza does not need a tomato topping. You can put ingredients directly onto the top of the pizza, such as marinated or sautéed artichokes, eggplants, zucchinis, asparagus, onion slices, rosemary, or a selection of white cheeses, all drizzled with a fragrant olive oil and a good sprinkling of garlic.

Pizza Toppings

ARTICHOKE AND GOAT'S CHEESE

To the basic pizza, add a handful of sliced, blanched, or marinated (from a deli or a jar) artichoke hearts, and a few ounces of crumbled goat's cheese with a sprinkling of shredded mozzarella or fontina. Drizzle with olive oil and bake as directed.

MERGUEZ AND EGGPLANT PIZZA

In Provence, instead of the ice-cream truck making the rounds on a summer's night, the pizza man comes instead. These little trucks have wood ovens installed in them, and turn out amazingly good pizzas. My favorite is merguez and eggplant: Top the basic tomato-and-cheese pizza from above with a scattering of lightly olive-oil–browned eggplant pieces and bits of merguez sausages. Bake until the sausages are browned, and serve.

PIZZA NAPOLITANO

Make a pizza base using half the quantity of pizza dough (pg. 114). Spread with a thin layer of tomato paste; one red onion, thinly sliced; several very ripe, flavorful tomatoes, diced (canned are fine); five or so anchovies, diced; five to ten black oil-cured olives (pitted and halved), two rounds of fresh mozzarella, thinly sliced; one teaspoon of capers, and two garlic cloves, finely chopped. Drizzle with three or four tablespoons of olive oil, sprinkle with one tablespoon of oregano leaves, and bake. Serve with freshly shredded Parmesan and a sprinkling of red, hot-pepper flakes, as desired.

OLIVE, MUSHROOM, AND RICOTTA CALZONE

Prepare the pizza dough and roll it out thinly. Scatter the bottom with thinly sliced mushrooms, scoops of ricotta cheese, slivers of black, oil-cured olives, a drizzle of olive oil, and a sprinkling of fresh thyme or rosemary. Fold over to enclose, seal the edges well, then spread tomato paste, diced tomatoes, cheese, and more olive oil over the top of this. Bake as directed above.

I sometimes add diced, green bell pepper to this, or a bit of browned sausage, and always eat with sprinkled Parmesan and red, hot-pepper flakes, and a few more olives on the side.

CALIFORNIA CLASSIC CALZONE OF GOAT'S CHEESE AND GREEN CHILE

Prepare the calzone as above, but fill the dough with goat's cheese, chopped garlic, chopped cilantro and green chile pepper. Top with tomato paste, and bake as above.

◀ *Pizza Napolitano*

pizzas, sandwiches, and savories

Garbanzo Crêpe

socca

Socca—the cry in the marketplace of "tout chaud"—and the little boys on bicycles bring fresh socca from the bakery around the corner, and all gather to have a slab of this freshly made treat. Socca is only a thin batter of garbanzo flour and olive oil, baked into a large, thin crêpe. A sprinkling of either rosemary or cumin adds extra flavor, but black pepper is the only traditional seasoning, though in Nice you'll find socca made in wood-burning ovens so that it has an elusive, smoky scent. For that aroma at home, I often make it on the grill.

SERVES 4

- 2 cups garbanzo flour (also known as *besan* or gram flour)
- 1 tsp salt
- 3–5 Tbsp extra-virgin olive oil, plus extra to cook socca in
- 1–1½ cups cold water, or more, enough to make a thick-enough batter
- Freshly ground black pepper
- Pinch of cumin
- Tiny plate of Niçoise olives, to serve

Preparation: 15 minutes

Cooking time: 15–20 minutes

❶ Stir the garbanzo flour and salt together. Mix the olive oil with the water, and stir it into the garbanzo flour. It will form a smoothish batter, but will still have a few lumps. Press it through a strainer to rid it of lumps, or put it in the blender. Season with cumin and pepper.

❷ Heat a tablespoon of the olive oil in a heavy skillet, and when it is very hot, ladle in enough of the batter as you would for making crêpes, and swirl it around in the same way. Cook over a medium heat to cook through the bottom, then place it in the broiler close to the flame to cook the top. Serve right away; for the full flavor of Provence, have a glass of chilled rosé and a little plate of tiny Niçoise olives along with it.

Salsas, sauces, and condiments—those little, strongly-flavored concoctions that imprint all they touch with their own distinctive character. Olive oil on its own is a good condiment, as used throughout the Mediterranean. Sometimes mixed with lemon and herbs, sometimes with sea water, and often with nothing at all, served in a simple cruet for drizzling over soups or risottos, grilled fish, or a crisp salad. Olive oil to dip bread into, to pour over beans, to whisk into mayonnaise, or to simmer into mojo.

Sometimes it is the olive itself—feisty, pungent, infinitely varied, it is the basis of many sauces such as tapenade, or even mustard. The following is a selection of favorite condiments with an olive flavor and goodness that will transform and enhance any dish.

Pistou

pounded paste of basil, garlic, and olive oil

Unlike pesto, pistou does not have a thickening made with crushed nuts, but like pesto, it is often enriched with grated, sharp, Parmesan cheese. Stir pistou into pasta, steamed vegetables, or a Mediterranean-style vegetable soup.

Pistou flavors vary, between using more or less garlic for pungency, extra basil for freshness and fragrance, and olive oil for a smooth, rich texture.

SERVES 4

• 3 garlic cloves, peeled
• Several handfuls of fresh, sweet-basil leaves, torn coarsely
• 5 Tbsp extra-virgin olive oil or more, as needed
• 6 Tbsp shredded Parmesan, or to taste

Preparation: 10–15 minutes

❶ Crush the garlic cloves with a mortar and pestle, then transfer it to a food processor or blender, and continue crushing. Add the basil, then slowly add the olive oil, working the mixture in until it forms a smooth paste. Add enough olive oil for it to be smooth and oily, then stir in the cheese.

❷ Store the pistou in a bowl or jar with a layer of olive oil over the top, for no longer than two weeks. If you wish to store it longer, you can freeze it for up to four months, but omit the cheese when freezing to ensure freshness.

Mustard Vinaigrette

vinaigrette à la moutarde

Whenever Esther and I get together, we inevitably make salad, and we always end up making the following mustard-based dressing. If you like mustard, this is the dressing for you. Try it on a mixture of sturdy and delicate leaves, with a good scattering of fresh herbs, such as tarragon, chervil, chives, and shallots.

SERVES 4–6

• 1–3 garlic cloves, chopped fine
• 2 Tbsp French mustard, such as Maille or Dijon, strong or mild, or use a whole-grain mustard
• 1–3 tsp vinegar, preferably balsamic, or a mixture of balsamic and raspberry

• 3 tsp extra-virgin olive oil, or more if needed

Preparation: 15 minutes

Combine the garlic with the mustard and vinegar, then slowly add the olive oil, stirring well to combine it into a thick, yellow mixture. It should taste strongly of mustard, and should cling to the leaves of a salad with its glistening, olive-oil goodness.

▲ *Pistou*

119

Garlic-and-Chili Mayo

rouille

Rouille is a rust-colored, garlic-and-chili mayonnaise from the south of France. This one, however, has a southwestern U.S. influence, and it is both simple to prepare and utterly delectable. Serve classically, with any boiled fish or fish soup, or slather onto grilled eggplants, or serve with roast pork or grilled fish.

SERVES 4

- 3–4 garlic cloves, crushed thoroughly with a mortar and pestle
- 1 tsp mild chili powder
- 1 tsp paprika
- 1/8–1/4 tsp cumin seed
- 3–4 heaped Tbsp mayonnaise
- Salt and cayenne pepper to taste
- Juice of 1/4 lemon, or to taste
- 3 Tbsp extra-virgin olive oil
- 1 Tbsp chopped, fresh cilantro

Preparation: 15 minutes

Chilling time: at least 1 hour

❶ Combine the garlic with the chili, paprika, and cumin seed, then stir it into the mayonnaise. Stir in the salt, cayenne pepper, and lemon juice, then slowly work in the olive oil, letting the sauce absorb it all before you add a little more. If the sauce can absorb more than three tablespoons, by all means add it. The richer the olive flavor, the more succulent the sauce.

❷ Taste for seasoning, then stir in the cilantro. Chill before serving.

Black-Olive Aioli

This luscious spread or dip is startlingly purplish-black in hue, with a gloriously rich, olive flavor. Spread lavishly onto crusty ciabatta bread, then layer with roasted bell peppers, salami, and arugula, or serve dabs of the aioli with chicken breasts, tomatoes, roasted red peppers, and grilled asparagus.

SERVES 4

- 3 garlic cloves, chopped
- 2 Tbsp black-olive paste
- 2 Tbsp extra-virgin olive oil
- 3–5 Tbsp mayonnaise, or as needed
- 1 Tbsp chopped fresh basil

Preparation: 5–10 minutes

Combine the garlic with the olive paste, and add it to the mayonnaise in spoonfuls, until the mayonnaise turns a purplish color and tastes richly of olives. Take care, as too much olive paste can curdle the mayonnaise. Stir in the rosemary or basil, and chill until ready to serve.

salsas, sauces, and condiments

Cuban Garlic-citrus Sauce

mojo

Mojo is Cuba's national table sauce, and like many Cuban foods, it has found a new home in southern Florida, where the traditional sour or Seville orange has been adapted to make all manner of citrus juices.

Here, I use a combination of tangerine and lime juice for a fragrant, sunshine-packed sauce. The balance of sweet, tart citrus, musky cumin, and fragrant garlic-flavored olive oil is delicious on almost anything, especially something cooked over a grill, such as lamb or sweet potatoes.

SERVES 4
- 3–4 Tbsp extra-virgin olive oil
- 8 garlic cloves, thinly sliced
- 1 cup tangerine juice
- ¼ cup lime juice
- ½ tsp ground cumin
- Salt and black pepper, to taste

Preparation: 5–10 minutes

Cooking time: 5 minutes

❶ Gently heat the olive oil with the sliced garlic until the garlic turns light golden (about 30 seconds), then add the remaining ingredients and remove from the heat.

❷ Let cool to room temperature, and taste for seasoning. This keeps and stays delicious for about three days in the refrigerator.

Tarragon-Infused Olive Oil

A few drops of this sprinkled around a plate creates a party-like effect, with its leafy, green hue and jolt of tarragon flavor. Note: You must follow the recipe instructions carefully as home-prepared infused oils can cause botulism (pg. 13).

MAKES ABOUT ½ CUP
- Several handfuls of fresh (a small bunch) or 3–4 Tbsp frozen tarragon
- ½ cup extra-virgin olive oil

Preparation: 10 minutes

Infusing time: 2 hours

❶ Blanch the fresh tarragon, if using, or half-defrost the frozen tarragon. If using fresh, drop it into ice water before blanching, then squeeze out the excess water. The ice water will preserve the bright color and crisp texture of the herb. Chop coarsely.

❷ Combine the herbs with the extra-virgin olive oil in a blender and purée, off and on if you need to, until it is a smooth mixture of bright-green, thickly herbed oil. Pour into a container, cover, and leave to infuse in the refrigerator for a maximum of two hours.

❸ Sift through a strainer or cheesecloth. This oil must be kept in the refrigerator and consumed within 12 hours.

Chile-and-Olive Oil Salsa

salsa peruana aji

Aji is the Peruvian hot pepper, yellow in color and more sweet than hot, though there are much spicier varieties. This salsa has a delicious flavor, and is great spooned onto grilled chicken, into black-bean soup, over a rare steak cooked over the hot coals, or a bowl of plain, steamed rice or pasta. It's also very nice drizzled onto peeled and blanched baby broad-beans, along with a shower of thinly shaved Parmesan or Romano cheese.

When choosing chiles for this dish, balance the heat and sweet-pepper flavor; for this you must know the heat of your chiles, and there is only one way to know this— by taste—carefully and judiciously proceed.

MAKES ABOUT ¾ CUP
- ½ cup extra-virgin olive oil
- 1 red bell pepper, diced
- 2–3 fresh red chiles, preferably large, somewhat-mild ones (if they are particularly mild, omit the red bell pepper and use only chiles), thinly sliced
- 5 garlic cloves, chopped
- 3 Tbsp lemon juice
- ½ tsp salt, or to taste
- Pinch of oregano, to taste

Preparation: 5–10 minutes

Cooking time: 15 minutes

❶ In a skillet, heat the olive oil with the pepper, chiles, and garlic, cooking and stirring for a few moments. Add the lemon juice, water, salt, and oregano, and simmer for about 10 minutes.

❷ Taste for seasoning, and serve either warm or cool.

Andalusian Tomato Vinaigrette

Tomato vinaigrette is delicious on so many of the sun-drenched salads of Andalusia. I especially like it tossed onto garbanzos, or with grilled artichokes (pg. 60).

SERVES 4
- 2–3 large garlic cloves
- Salt to taste
- 3 ripe tomatoes (canned are fine)
- 1 Tbsp balsamic vinegar
- 1 Tbsp sherry vinegar
- Pinch of oregano
- About 3 fl oz extra-virgin olive oil

Preparation: 10 minutes

❶ Using a mortar and pestle, grind the garlic with the salt until it forms a paste. Dice the tomatoes and work into the mixture. Add the vinegar and oregano, then slowly add the olive oil until it forms a vinaigrette.

▲ *Chile-and-Olive Oil Salsa*

Greek-style Garlic Sauce

skortalia

This Greek-style garlic sauce is as much about olive oil as it is about garlic. It is actually a mayonnaise, but does not use eggs to emulsify it. Skortalia is eaten with all sorts of vegetables—if you order fried zucchinis in Athens, it will usually come with this rich mixture. Crisp fish—either fried or barbecued—are classically served with skortalia, as are boiled beets. Hard-cooked eggs are also lovely with this sauce.

SERVES 4

- 2–3 large garlic cloves
- Large pinch of salt
- Juice of ½ lemon, or more to taste
- 1 small, freshly boiled, and tender, peeled potato, broken into several pieces, plus a few Tbsp of its cooking water
- ½ cup extra-virgin olive oil, or as needed

Preparation: 15–20 minutes

Cooking time: 15–20 minutes

❶ Using a mortar and pestle, crush the garlic with the salt until it forms a paste, then work in the lemon juice and mix until it is smooth and creamy.

❷ Work in the potato and its cooking water, then slowly work in the olive oil, starting with a tablespoon or so, and repeat until the sauce is rich and delicious. After a while, you can add a slightly larger amount at a time, and stir it in with a fork or spoon.

salsas, sauces, and condiments

Olive oil is a great cooking medium for desserts as well as savory foods. In the Middle East, for instance, there are numerous pastries of deep-fried yeast doughs, served with syrups or honey. My favorites are the Greek loukamades, which are basically tiny, fluffy doughnuts drizzled with honey, cinnamon, and sesame seeds. Pure olive oil, with perhaps an added hint of extra-virgin olive oil, fries these to a delicate, crispy, golden color. Ordinary doughnuts are delicious cooked in olive oil; simply sprinkle them with sugar and cinnamon, and brew a nice cup of strong black coffee to have with them.

You can also use olive oil for cakes, in place of other oils, butter, or margarine. Olive oil makes for a moist cake, and the olive flavor will only be as strong as the olive oil itself. Even with strong oils, the olive scent is elusive, and smells like something pleasant but unidentifiable. The following are a couple of moist, olive-oil–scented cakes.

a few desserts

Muscat-and-Peach

olive-oil-scented sponge cake

This makes a plain, fragrant sponge cake, perfect for nibbling on with a glass of late-afternoon muscat on a Mediterranean sojourn ... it is at its best the next day, when it has had a chance to firm up and become mellow. I usually serve the cake with a demitasse of black coffee or espresso, and a bowl of sugared, sliced, fresh peaches, doused with a little more wine.

Though I call for a mild olive oil, the truth is that I use whatever I have to hand, and have made this using strong, full-flavored, extra-virgin olive oils, with lovely results.

SERVES 6

- 4 eggs, separated
- Pinch of salt
- 1 cup raw brown sugar
- 1 ripe peach, diced
- ½ banana, diced
- 1 tsp rose water
- ⅛–¼ tsp almond extract, or to taste (optional)
- 1¼ cups all-purpose flour
- ½ cup muscat or other white wine
- ½ cup mild, extra-virgin or pure olive oil

Preparation: 30 minutes

Cooking time: 40 minutes

❶ Combine the egg whites with the salt, and set aside. Whip the egg yolks and mix in with the sugar, beating until they are creamy in consistency; then add the peach, banana, rose water, and almond extract, and beat for a few moments longer or until frothy.

❷ Sift the flour into the liquid and stir gently until well blended, then stir in the wine and olive oil.

❸ Whisk the egg whites until they form firm peaks, then fold about a third into the batter to lighten it, and fold the rest in. Fold in the wine and olive oil.

❹ Grease and flour a cake pan, and place a piece of buttered parchment the same size and shape of the pan bottom. Pour in the batter, then place in a preheated oven at 375°F, and bake for about 20 minutes.

❺ Lower the heat to 325°F, and bake for another 20 minutes or so, then remove from the oven and leave to cool. When the cake is halfway cool, loosen the sides with a knife, then invert gently.

Gooey-Chewy, Double-Ginger Cake

The sides of this cake rise up into chewy, almost-crisp edges, while the middle slumps into an almost-gooey confection. Chunks of chopped, fresh ginger add zing, and the olive oil keeps it moist and hearty.

SERVES 6

- ¾ cup raw brown sugar
- ¾ cup superfine granulated sugar
- ¾ cup corn syrup
- 1–2 Tbsp fresh gingerroot, chopped coarsely
- Pinch of salt
- ½ cup dark-brown granulated or molasses sugar
- ¼ cup coffee (I use cold, leftover coffee from breakfast)
- ¼ cup skimmed milk
- 3 fl oz extra-virgin or pure olive oil
- ¾–1 tsp ground ginger
- ¼ tsp ground cinnamon
- Large pinch or two of ground cloves, or to taste
- 2 cups all-purpose flour
- 1 egg, lightly beaten
- ¾–1 tsp vanilla flavoring and/or ¼ tsp vanilla extract
- Few drops of vinegar

Preparation: 30 minutes

Cooking time: 1 hour

❶ Combine the raw brown sugar, white granulated sugar, corn syrup, fresh gingerroot, salt, and dark-brown granulated or molasses sugar in a saucepan, and heat over a medium-low flame until the sugar and syrup melt together. Let it cool a bit, then stir in the coffee, milk, olive oil, ground ginger, cinnamon, and cloves. Slowly add the flour, egg, vanilla, and vinegar.

❷ Pour into a round, oiled, and floured baking pan, and bake at 350°F for about an hour or until the sides start to pull away from the pan. Test the inside, if it seems too runny, turn off the oven and let the cake sit in the oven as it cools, for about 15–20 minutes or until the middle seems chewy, but settled.

❸ Remove from the oven, and let cake cool.

Index